D0065265

CHEMICAL METHODS OF SILICATE ANALYSIS

—A HANDBOOK

CHEMICAL METHODS OF SILICATE ANALYSIS

—A HANDBOOK

by

H. BENNETT, B.Sc., A.R.I.C., F.I.Ceram.

Head of the Analysis Department, the British Ceramic Research Association

and

R. A. REED, M.Phil., A.R.I.C., A.I.Ceram.

Scientific Officer, the British Ceramic Research Association

1971

Published for the

BRITISH CERAMIC RESEARCH ASSOCIATION

by

ACADEMIC PRESS, LONDON AND NEW YORK

ACADEMIC PRESS INC. (LONDON) LTD.
24–28 Oval Road
London, NW1 7DX

U.S. Edition published by
ACADEMIC PRESS INC.
111 Fifth Avenue,
New York, New York 10003

Library of Congress Catalog Card Number: 75–170749

ISBN: 0–12–088740–1

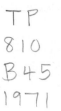
PRINTED IN GREAT BRITAIN BY
C. TINLING & CO. LTD, PRESCOT AND LONDON

FOREWORD

by

N. F. ASTBURY, C.B.E., M.A., Sc.D., C.Eng., F.I.E.E., F.Inst.P., F.Inst.F., F.I.Ceram.

Director of Research, The British Ceramic Research Association

In a chemical world increasingly furnished with ingenious "black boxes", contrived to take the human drudgery out of analysis, one is apt to overlook the fact that these devices can function only if in support there is a soundly based practice of reliable direct chemical methods. In 1958 the Research Association published what was in effect a handbook of such methods—Bennett & Hawley's "Methods of Silicate Analysis".

Continued research at the British Ceramic Research Association in this otherwise almost neglected field of analytical research prompted a second edition in 1965, and it is clear that the methods developed by the authors and their colleagues have been conspicuously successful in their chosen field, and have in fact formed the basis of various British Standards.

The present book, like the 1958 edition, is simply a handbook of methods, and discussion of procedures, which was included in the 1965 edition, has been omitted.

The authors feel that a limit has very nearly been reached in the speed and accuracy of the methods described, and indeed the pressure on purely chemical procedures has now been greatly eased by instrumentation. Thus it is likely that this book may have more permanence than its two forerunners, which were published when thinking about analysis was changing pretty rapidly.

It is hoped that this new volume will prove as useful a "bench book" as its predecessors and that it will meet adequately the needs of both the referee analyst and of the small routine laboratory.

Finally, I record that W. G. Hawley, co-author of the 1958 and 1965 editions, has retired, and we are pleased to see the name of Richard Reed on the title page of this book.

PREFACE

The first edition of "Methods of Silicate Analysis" by H. Bennett and W. G. Hawley appeared in 1958, the second in 1965. Such has been the pace of the advance in ceramic analysis that even the second edition contains much that is outdated. In the first edition the classical method of analysis, based on a quantitative exploitation of the qualitative analysis tables, was paramount; in the second this was replaced by other methods for the analysis of most materials. The classical method was recommended only where the type of material had not at that time been investigated or where the material did not conform to type and therefore presented problems in applying methods designed for a specific range of chemical composition. The newer methods described in the second edition were based on the principle of a single dehydration technique for silica and the use, wherever possible, of separations by solvent extraction followed by determination with EDTA. The present volume continues this trend towards newer methods—only bone ashes and chrome-bearing materials now have procedures which are essentially classical. Even for chrome-bearing materials, although the classical method is given, the alternative method has proved more reliable.

The discovery that polyethylene oxide can coagulate gelatinous silica, such as can be obtained by the acid dissolution of an alkaline fusion or a direct acid dissolution of a sample, and permit the determination by colorimetric means of the amount of silica left in solution, has resulted in a new generation of methods. Although the single dehydration has been included in the techniques described in this volume, the authors consider that coagulation is the preferred technique. It is not only faster, eliminating a break in the continuity of the analysis while the dehydration proceeds, but also tends to give more accurate answers since the residues after hydrofluoric acid treatment tend to be smaller. This is particularly so in the analysis of zircon, where the very low levels of hydrolysis of zirconium salts during coagulation as compared to the very high proportion of zirconium precipitated during dehydration, has turned a very difficult analysis into a relatively simple exercise.

It is difficult to visualize developments in traditional methods continuing at the pace of the last few years. Most types of material which are being regularly analysed by the industry have now been scrutinized and can be handled by relatively modern methods, so that the urgency of the demand for a reliable standard method which is reasonably fast has decreased. Savings in time to be effected by any future changes are likely to be marginal and thus little more than a watching brief is likely to be held in this sector. Apart from this watching brief, the emphasis in terms of analysis has turned towards instrumental methods, in particular X-ray fluorescence. Thus the methods described in this book are of limited interest. In the authors' opinion these methods serve two main purposes in providing the analyses on which the instrumental analyst can base his comparison as well as simple techniques for

vii

those laboratories which do not perform enough analyses to justify the expensive equipment.

This volume is not the place to discuss the merits of "wet" analysis *vis-à-vis* instrumental analysis; it is sufficient that the authors are convinced that for many years there will remain a need for the type of methods described herein. The scope of the book is deliberately restricted to this traditional approach; the only equipment demanded by the procedures being a pH meter, a flame photometer and a spectrophotometer.

The problems posed by the need to analyse ceramic materials have been solved in other less traditional ways; X-ray fluorescence, direct-reading spectrometry and atomic absorption spectrophotometry are all widely employed. If the first two are used, a large "through-put" of samples is necessary to justify the purchase of the equipment; in the case of the last a considerable amount of research is still being conducted. Individual laboratories have developed procedures to suit their needs, but there is, as yet, no agreement about the correct approach. The authors have, in any event, reservations concerning the use of A.A.S. for the determination of major contents, particularly for elements as difficult to determine as silicon and aluminium. Determinations such as calcium in magnesite and chrome-bearing materials seem ideal for the application of A.A.S., and the technique is likely to form the basis of a number of standard methods in due course. On the other hand, determinations of elements like iron, manganese and chromium bring the atomic absorption technique into direct competition with spectrophotometry and the balance of advantage does not necessarily lie with the newer approach. There is, at present, too much interest in using the technique as a universal panacea but, as has already happened with flame photometry, it will doubtless find its correct place as an analytical tool within a few years.

The main emphasis of analytical research in ceramics is now moving towards a wider approach than that of improving methods of analysis for the traditional elements in various bulk ceramics. The function of elements not normally determined in ceramic technology is increasing in significance, the discovery of the importance of the presence of boron in magnesite refractories drew attention to this. In addition to the determination of so-called "trace elements", there is a growing interest in elucidating problems of structure so that it is becoming increasingly difficult to draw a clear line to indicate where analysis ends and structure investigation starts.

In selecting analytical procedures for inclusion in this book, the authors have adopted a similar policy to that used in the two editions by Bennett & Hawley, namely that all the methods described are personally familiar to the authors and are known to give reliable results when properly applied to the materials for which they have been designed. Therefore, although the reader may not find his favourite method described, he should at least find an acceptable alternative. More than in either of the previous volumes the methods described are essentially developed from the same basic conceptions.

In order to make each procedure as self-contained and intelligible as possible, a complete chapter is devoted to each material and the full methods of analysis are detailed. Reagents are incorporated in each chapter, but these are only dealt with in detail when their preparation contains a complication. It is assumed that an analyst is sufficiently intelligent to make a liquid/liquid dilution from the description $(1 + n)$ and a solid/liquid dilution from the description (ng/litre). By this means it is hoped that the method will be easier to follow in that fewer references back will be needed to identify solutions. ISO standard methods, for example, appear to the authors to be so highly codified in the interests of "clarity" as to be almost unintelligible.

The choice of methods for inclusion in this volume has been so wide that on this occasion some have been eliminated. The rapid and spectrophotometric methods for high-silica and aluminosilicate materials have just escaped the axe, but simply on the grounds that in some ceramic laboratories they have not yet been replaced; they should, however, be regarded as obsolescent. Improved methods for materials such as magnesites and chrome-bearing materials have enabled earlier, more tentative methods to be discarded.

Since the compilation of most of these methods, DCTA has been proved to be an acceptable alternative to EDTA for the volumetric determination of alumina. It has several advantages, among which are greater speed and the non-interference of chromium. The method has been described in a separate chapter (Chapter 17) but it may preferentially be used to replace the EDTA procedure in any of the appropriate methods.

From the distribution of the earlier volumes it is clear that the first three chapters, on general considerations and equipment, have been of value to the smaller laboratories and to students. They have therefore been retained, although it is realized that they will not interest all readers.

A handbook of this size must restrict its coverage—this is concerned solely with traditional methods as a matter of choice. Instrumental methods form a separate topic.

July 1971
                                        H. BENNETT
                                        R. A. REED

ACKNOWLEDGEMENTS

The authors are grateful to all those whose work has been used in the preparation of this book. It is impossible to acknowledge all the authors from whose papers information or parts of methods have been taken. However, a limited number of references have been made in the text to significant contributions, and the following publications have been particularly useful:

HILLEBRAND, W. F., LUNDELL, G. E. F., BRIGHT, H. A., and HOFFMAN, J. I. (1953). "Applied Inorganic Analysis". Wiley, New York.

MELLOR, J. W., and THOMPSON, H. V. (1938). "A Treatise on Quantitative Inorganic Analysis", Griffin, London.

The authors also wish to acknowledge their debt to the various committees associated with chemical analysis at the British Ceramic Research Association, to the co-author of the previous volumes, Mr. W. G. Hawley, much of whose work lives on in this book, and finally to Mr. K. H. Speed, Mrs. M. Horton and Miss V. A. Knapper for their help in its publication.

Extracts from B.S. 784:1953, Methods of test for chemical stoneware; B. S. 3921:1964, Methods of testing clay building bricks; B.S. 1902, Methods of testing refractory materials; Part 2A:1964, Chemical analysis of high-silica and aluminosilicate materials; Part 2B:1967, Chemical analysis of aluminous materials; Part 2C:1967, Chemical analysis of chrome-bearing materials; Part 2D:1969, The determination of silica in high-silica, alumino-silicates and aluminous materials by coagulation; Part 2E:1970, The chemical analysis of magnesites and dolomites, are reproduced by permission of the British Standards Institution, 2 Park Street, London W.1., from whom copies of the complete standards may be purchased.

CONTENTS

PART I

CONSIDERATIONS AND EQUIPMENT

PART II

METHODS OF ANALYSIS

Part I. Considerations and Equipment

CHAPTER 1

PRELIMINARY CONSIDERATIONS

Unlike many fields of analysis where the percentage of a given element present is recorded, the silicate and ceramic analyst reports the percentage of the oxide. This convention is not merely an empirical convenience to save calculating or even determining the oxygen content of the sample, but represents a truer picture of the constitution of the material. Kaolin or china clay approximates to the formula $Al_2O_3.2SiO_2.2H_2O$, so that if the results are expressed as percentages of SiO_2 and Al_2O_3 a more useful picture emerges than if the Si, Al and O contents are reported. The origin of the convention probably lay in the method of classical analysis wherein most of the constituents were actually weighed as oxides.

A further convention to be noted is that of the full or complete analysis. The industrial analyst is required to produce analyses which are relevant to the problems of production or the behaviour of the finished product and the only constituents which are therefore of interest are those which have a bearing on these. When ordinary high-silica materials or aluminosilicates are being analysed, for instance, silica bricks, ganisters, flints, sands, clays, firebricks, pottery bodies and sillimanites, the conventional full or complete analysis would normally comprise the determination of silica, alumina, titania, ferric oxide, alkali oxides, lime and magnesia.

The only alkali oxides reported were those of sodium and potassium, but the advent of the flame photometer has rendered the determination of lithia easy and this constituent is now often included. Occasionally manganese oxide and sulphur trioxide are determined, and on very rare occasions a determination of ferrous oxide is called for. Determination of hygroscopic water, combined water, carbonate and carbonaceous matter are rarely required; in normal practice these, with the exception of hygroscopic water, are included in the loss on ignition. The results are customarily reported as follows:

Constituent	Percentage
SiO_2	
TiO_2	
Al_2O_3	
Fe_2O_3	
Mn_3O_4	
CaO	
MgO	
Na_2O	
K_2O	
Li_2O	
Loss on ignition	
Other oxides	

B

1

The assumption is made that titania functions as an acidic oxide, hence its position next to silica. The other constituents then follow in the normal order of determination in a classical analysis with the exception of the alkali oxides for which a separate solution is prepared, and the loss on ignition, often the first determination.

Although the ceramic analyst is content to carry out the normal complete analysis he should be aware that it ignores the possible presence of many other constituents. Zirconium, chromium, vanadium, phosphorus and other elements are often found, but in very small amounts and, what is probably more important, their presence is of little consequence in practice. There is, of course, a fundamental distinction between ceramic analysis and geological or mineralogical chemical analysis. In geological analysis the presence of trace elements may at times be of major importance.

Methods designed specifically for routine control analysis are few, and this area is being taken over by instrumental methods. Many laboratories use other methods for control of their particular materials, often based on the classical method modified to increase speed. None of these methods is included in this book since it is very difficult to determine just how much accuracy has been lost as a result of the speeding up. Only methods designed for rapid routine analysis to which a precise figure of accuracy can be attached have been included.

It should be appreciated that the ceramic analyst's definition of "rapid" may be radically different from that of analysts employed in other fields, for example, the iron and steel industry. In the latter industry no method would be described as rapid unless the final result was available in less than half an hour and often times of the order of 10 min are needed. This limitation is almost unattainable in ceramic analysis except in a few isolated cases. Before routine methods were introduced there was no alternative to analysis by the classical methods which took from 3–10 days depending on the accuracy required and the material to be analysed. It is against this background and the long-standing tradition of very slow analysis that the term "rapid" should be viewed.

In addition, almost all ceramic materials need a fusion technique to bring them into solution, a process which is in itself comparatively time-consuming. Also many of the important rapid methods of analysis relate to the determination of major constituents, and therefore, to attain a satisfactory standard for even routine analysis, a relatively high degree of accuracy must be maintained. Errors greater than 0.5–1% of the amount present are rarely permissible when dealing with this type of determination. For these reasons it usually follows that extremely fast methods have to be rejected as insufficiently accurate, so to obtain the accuracy required more processes have to be introduced and more precautions taken, thus greatly prolonging the determination.

The present volume contains methods which demonstrate that rapid control analyses can be carried out by standard methods; for example the coagulation technique, by ignoring the residue after treating the silica with hydrofluoric acid, can be completed in a day. The standard method is now competitive in speed with that used for control purposes, except in a few special circumstances. At the other end of the scale, namely in further increased speed even at the loss of some accuracy, the "wet" control method is being replaced by instrumental techniques.

CHAPTER 2

PRACTICAL CONSIDERATIONS

A number of readers of this book may not have had the opportunity of training at an advanced level. This chapter and the next have been included to ensure that these readers pay due attention to important details which can make or mar the performance of a successful analysis. It is appreciated that much of the following will be of little or no service to the skilled and experienced worker.

BULK SAMPLING

The whole subject of sampling is fraught with difficulty and it is by no means easy to suggest even guiding principles, to say nothing of laying down fixed rules. It might even be said that sampling is a subject which has been carefully ignored up to the present, probably because of the obvious complexities.

The selection of a bulk sample from a clay mine or pit, or from a stockpile, is a very difficult task if the sample is intended to represent the bulk closely, and unless this is true the whole object of sampling is missed. This is not the place to discuss sampling techniques, the analyst's work usually starting at a later point. Much work has been done on the sampling of coal and it is likely that the general principles will be relevant; a good description of these and also of the pitfalls has been published.*

It is sufficient to point out that if the sample is not truly representative of the bulk, then the whole of the subsequent work is so much waste of time and the results can even be misleading.

PREPARATION OF SAMPLES

Generally speaking this is where the work of the analyst really begins, since from the bulk sample it is necessary to prepare a few grams of powdered material which is chemically identical. This process can be in error for one of two reasons; firstly, because the actual process of taking a small portion of the sample has been incorrectly performed, resulting in it being not representative of the whole, and secondly, because during the process of grinding the sample of powder, contamination has been introduced.

Errors for the first reason can in great measure be overcome by adopting correct procedure in reducing the sample bulk. Again it is beyond the scope of this book to go into the whole of this question in detail, reference once more being made to the above.* The operation can be performed manually with success if the coning and quartering processes are carried out so as to avoid bias and segregation and to ensure that the particle size at any stage of the

---

*WILSON, C. L., and WILSON, D. W. (1960). "Comprehensive Analytical Chemistry", Vol. IA, p. 36 *et seq.*, Elsevier, London.

reduction is appropriate to the weight of material handled. It is important to realize that errors from this source cannot be eliminated completely but provided that sufficient care is taken, the errors can be reduced to a tolerable level. Various forms of mechanical sample splitter may also be used; these can reduce some of the pitfalls of manual operation.

It is equally impossible to eliminate errors arising from contamination during grinding. In theory, at least, any contact of the material with any surface whatsoever will introduce some contamination, but in practice the level of contamination introduced in this way can be ignored. It should be remembered, however, that unless the whole of the material is soft or extremely friable some contamination will be introduced. The degree of contamination will clearly depend on the medium used for reduction of the particle size and on the hardness of the material being ground. It is possible to reduce some raw clays in a porcelain mortar without noticeable contamination; on the other hand as much as 5% of silica can be introduced into a hard-alumina sample by grinding it in an agate mortar. Certain rules are obvious; the mortar chosen for reducing the particle size should, where possible, be appreciably harder than the material to be ground. This is, of course, a counsel of perfection since many fired ceramic materials are almost as hard and in some cases harder than any available mortar. Contamination in these cases is inevitable and it is only possible to "make the best of a bad job". For instance, with a hard aluminous material, it is often necessary to prepare two samples, one ground in, say, an alumina mortar so as to give minimal contamination in respect both of total weight introduced and of added iron content. This sample may then be used for the determination of the "true" iron content. A second sample is prepared in an iron mortar, thus introducing only iron, some of which can be removed by means of a magnet (provided that the original sample was not magnetic). This sample can then be used for the main analysis, allowance being made for the iron introduced.

The introduction of boron carbide as a material for use in mortars goes some way to minimizing the problem. Contamination of an alumina sample by boron carbide appears to be at only the $0.1-0.2\%$ level. If mortars of a practical size could be produced at an economic price many comminution problems would be solved.

In general, less contamination is introduced if the sample is tamped in the mortar than if it is ground. Unfortunately, mortars such as those made of agate are liable to damage, for instance, by chipping of the pestle if the tamping technique is used.

DRYING OF SAMPLES

Prefired materials, after being ground and transferred to a sample tube (3×1 in. or 2×1 in.), should normally be dried at 110°C for at least 2 h, whereas for clays and similar materials the corresponding time is 4 h. It is advisable, unless there is good reason to the contrary, to dry samples for analysis overnight at 110°C. Special precautions have to be taken with samples which may take up carbon dioxide from the atmosphere; it is not advisable to dry such samples for longer than is strictly necessary.

WEIGHING

After drying, it is necessary to cool the sample in a good desiccator over a suitable drying agent. For most analytical work of the type described, silica gel is adequate and moreover it is self-indicating, a change in colour from blue to pink showing that the gel is becoming exhausted; heating it in an air oven for a few hours at 110°C effects regeneration.

It is advisable to standardize the time of cooling of crucibles and ignited precipitates as far as possible; the time should be no more than is needed for the contents of the crucible to reach room temperature, because of possible dangers due to the slow picking up of carbon dioxide and occasionally, with certain precipitates, moisture from the silica gel. Samples in tubes, silica and porcelain crucibles, or large platinum dishes (3 in. and above) normally take 30 min to cool, whereas it is often possible to weigh platinum crucibles after only 20 min.

The operation of weighing should be carried out as rapidly as possible to avoid errors due to the adsorption of moisture while the object is on the balance pan, and in this connection recent developments in balance design should be noted. The newer types of weight-loading balance including the fully weight-loading, single-pan balances, although expensive, enable a crucible and its contents to be weighed in a few seconds and, apart from the increase in accuracy because of this speed, the saving in "analyst time" soon offsets the higher price.

There are two schools of thought concerning the weighing of samples. The first weighs an amount of sample which approximates to, say, one gram, and then calculates all the results from the weight taken. The second prefers to weigh the sample to exactly 1 g within the permissible experimental error so as to obtain a direct conversion into percentages by multiplying by 100. There are arguments in favour of both; by weighing accurately an approximate amount the time during which the material is exposed to the atmosphere is minimized, reducing the chances of moisture adsorption; however, the calculation of the results is now complicated arithmetically. The second method simplifies the number of calculations, but if the original weighing is slow there is a danger of sufficient moisture adsorption to disturb the results. In our opinion, if one of the newer balances can be used the advantage now lies in weighing off the exact amount required, the speed being such that the danger of errors is greatly reduced.

TEMPERATURE OF IGNITION

It will be seen subsequently that different ignition temperatures are recommended for different precipitates. As a general rule, silica and alumina precipitates should be ignited at 1200°C whenever practicable. This high temperature is essential for accurate work. Firstly, it helps to "dead burn" the material, i.e. to reduce its hygroscopic tendencies, thus making the weighing more reliable and secondly, certain precipitates, notably R_2O_3, lose weight on re-ignition at 1200°C after ignition to constant weight at 1000°C. The importance of this loss in weight will be appreciated when it is realized that from our experience an R_2O_3 precipitate corresponding to an alumina content of more than about 60% can decrease in weight by the equivalent of about 1%. Thus for ordinary routine work on firebricks etc., ignition at 1000°C may be tolerable, but if accurate results are required then it is essential to carry out the ignition at 1200°C.

Care should be taken at all times to maintain an oxidizing atmosphere in the crucible otherwise products of reduction can seriously damage the platinum. Iron will stain the crucible and its removal is long and tedious. Phosphate precipitates can also affect the structure of the platinum if any reduction takes place. For these reasons the use of electric furnaces is to be recommended if at all possible.

BLANK DETERMINATIONS

Only in a few places in the text have blank determinations been specifically mentioned, but if the maximum degree of accuracy is required then it is clearly essential to carry out a blank determination exactly as in the analysis, but omitting the sample. Occasionally, e.g. in the determination of alumina by oxine, a known amount of alumina must be added to obtain a blank by difference as very small amounts of alumina are not quantitatively precipitated, in which case the blank becomes almost a control determination. It must be left to the individual analyst to consider his own problem in terms of accuracy and then to decide how best to evaluate a reliable blank, if necessary. For instance, it is rarely wise to carry out a blank fusion in a platinum crucible, for experience has shown that the fusion mixture attacks the platinum much more readily in the absence of a sample. The platinum thus introduced has to be taken through the analysis and may lead to the determination of incorrect blanks.

The blank determination is usually considered necessary because of the introduction into the analysis, from the reagents, of the element to be determined, or possibly other elements which may react under the same conditions. Adequate care in the storage of liquid reagents can often help to reduce this error; plastic containers are recommended, and in particular all alkaline solutions and EDTA solutions should be stored in polythene bottles.

ROUTINE CONTROL ANALYSIS

There are several methods described in this book that have been devised specifically for routine control work. It is important to realize that the potential errors quoted for these methods represent the potential of the method when correctly carried out. It is a mistake to think that as the level of accuracy demanded of the results is less, correspondingly less attention need be paid to detail. Rigid attention to detail is even more necessary here than might be the case with a classical analysis. The reason is that the methods have been made so simple and fast that most of the available tolerance is taken up by the method itself and little is left for the operator. The same remarks apply when accurate methods are being used for routine control work, the work should always be performed at the highest level of accuracy possible, minor "simplifications" should always be avoided as it is often very difficult to prescribe a tolerance on a "modified" method; errors may be unexpectedly high unless the whole process is kept under strict control. Continuous slipshod working on the pretext that the results are "good enough for what is required" leads very easily to an inability to do good work at other times. In our opinion it is essential at all times to carry out the analysis, no matter what the basic accuracy of the methods used, as though it were a referee analysis: the errors are then those of the method, plus the inevitable human error.

CHAPTER 3

APPARATUS AND EQUIPMENT

The choice of suitable equipment for a laboratory which specializes in ceramic analysis is becoming a progressively more difficult problem as the range of time-saving apparatus increases. The working requirements of the individual laboratory must be taken into account and assessment made of the additional cost of newer types of equipment, the frequency of its use and the value of the time so saved. This problem can only be decided by a chemist who is fully familiar with the relevant statistics in the laboratory concerned. The information contained in this chapter is intended to assist in drawing attention to such innovations as may be profitable and in coming to a correct decision.

Much of the following is elementary but no apology is made for its inclusion, since, in view of the present structure of the industry and its increasing interest in analytical control, most of the information will doubtless be of use in some quarters. Apart from discussing equipment and apparatus as such, this chapter provides guidance on various points of technique concerned with the use of laboratory equipment, both general and specific, so that the maximum value can be extracted from what is available.

Two general points emerge from any consideration of apparatus and equipment for a laboratory. First, whatever equipment is deemed to be necessary should in fact be available and should be of the highest quality; the purchase of a cheaper type of apparatus is nearly always false economy. Secondly, and possibly almost conversely, it is desirable for the analyst to restrain himself from demanding apparatus which looks attractive, but cannot be economically justified; for example, "automatic" apparatus is uneconomic unless it is clear that either the present position or the position in the foreseeable future, will make it a paying proposition. Lavish spending on apparatus which is not strictly necessary can have unfortunate repercussions when later something essential is wanted.

It is intended that this chapter should primarily be of practical value in the selection and employment of the types of equipment necessary for a routine control laboratory and capable, if necessary, of handling accurate ceramic analysis. If the laboratory is also to be used for analytical research the requirements may be far more extensive. The various sections of this chapter are not in a strictly logical order, but the equipment is arranged roughly in the order in which it would be used in the course of an accurate analysis of an aluminosilicate material, with a number of small random items grouped together at the end.

SAMPLE PREPARATION

In most smaller laboratories, samples will be received in amounts of a hundredweight or less, usually in lump form, either pieces of clay, pieces of brick, whole bricks or specimens of ware. The actual treatment as far as the crushing process is concerned will vary with the nature of the material. Normally there will be a succession of crushings and sample divisions until a few grams are available, ground to pass a 120-mesh B.S. test sieve, or even finer.

Very often the first particle size reduction can be carried out conveniently by using a laboratory-size jaw crusher. Modern designs of this machine which are easy to clean and are therefore rapid in operation are available. By passing the material through the machine more than once, a maximum particle size of about $\frac{1}{8}$ in. can be attained. The contamination resulting from this treatment is normally small, although it must be appreciated that it will occur; the bulk of the sample is sufficiently large to bring the small amount of iron introduced down to a very low level when expressed as a percentage of the material. Even so, certain samples with a very low iron content may undergo significant contamination and this point should be borne in mind.

Sample splitting can be performed by hand by the process of coning and quartering, or by means of a sample splitter or riffler (Figs 1 and 2). The cleaning of mechanical sample splitters after service takes time and often cancels out any advantage. When the bulk of the sample is high, riffles can be used with advantage. They are designed to handle fairly large amounts of material and may not give a fair split when the bulk of the sample is small.

If the material being handled is fairly homogeneous it is possible that the bulk can be reduced to a few ounces before any further reduction in particle size, but in many instances further treatment is needed. In some circumstances the use of a set of mechanical rolls can be justified, but more frequently fairly large mortars of iron or porcelain are needed. Contamination difficulties can play a significant part in the considerations at this and subsequent stages and it is not possible here to discuss all the combinations of mortar and technique which should be used in specific circumstances. Consideration should be given to the nature of the material, its hardness or friability and also to the degree of accuracy required from the analysis. It is possible to tolerate greater levels of contamination in partial analyses in some cases than it would be if the whole analysis was being carried out. For instance, if the alumina content of a silica brick is required it is normally possible to prepare the sample in mortars made of iron; the total amount of iron likely to be introduced will not materially affect a very low alumina content.

Mortars

A large range of mortars is now available and, although it is impossible to go into great detail about the application of each type, the following notes may provide a useful guide to the choice for any particular purpose.

Iron

Large iron mortars are useful for breaking down fairly large amounts of sample where the material is not too hard or where the iron content is not of vital consequence. Tamping is preferable to grinding as it introduces much less iron into the sample. The metal of the mortars is usually sufficiently pure for

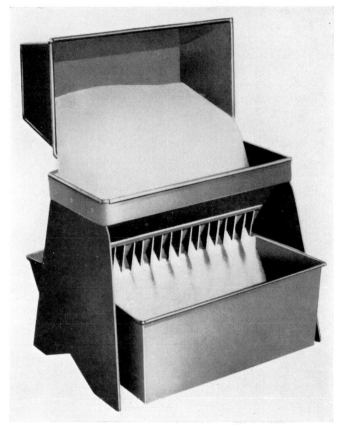

Fig. 1. Tyler Sample Splitter. Reproduced by courtesy of International Combustion Products Ltd.

contamination other than iron to be ignored. Much of the iron introduced can be removed by treating the gound sample with a magnet, provided always that the sample itself contains no magnetic materials.

Iron Percussion

This form of iron mortar, combined with the way it is used, tends to introduce considerably less iron into the sample than an ordinary iron mortar. As a result, it is possible to use this type of mortar for preparing samples considerably harder than could be tolerated in a large iron mortar, but apart from this the above remarks also apply. Only small amounts of sample can be dealt with, due to the restricted size of these mortars.

Stoneware

The sample must be very soft and friable. Only occasionally has this mortar an advantage over iron, and any contamination introduced cannot be

Fig. 2. Pascall Sample Splitter. Reproduced by courtesy of Pascall Engineering Co. Ltd.

removed and will be spread over many of the elements sought. It is desirable to restrict its use to previously ground material or to soft materials such as clays, and even here it is advisable to bear in mind that fireclays can contain nodules of hard minerals. It is possible to deal with fairly large amounts of sample.

Glazed porcelain or glass mortars are rarely suitable for the preparation of ceramic samples. The number of occasions on which they could be used would not normally justify their purchase.

Agate

As agate is a hard material it can be used for a fairly wide range of samples. It is, however, rather brittle and so should be used for grinding and not tamping. The degree of contamination is such that aluminosilicate refractories can be ground with safety, but if the alumina content is more than about 45% there is a danger that undue amounts of silica will be introduced. This can affect the results for alumina content appreciably, contaminations of up to 5% silica having been recorded on materials containing approximately 99% alumina. Agate also has the advantage that it is relatively pure and the contamination can be regarded as being only silica, a fact which is useful both when partial analyses are required or when two or more portions of the sample are being prepared for various parts of an analysis.

Alumina

In recent years mortars made of 99% or more alumina have become available; these are very hard and are in some ways complementary to agate. They are very convenient for grinding materials having alumina contents of 45–98%. Above 98% the degree of contamination is severe and as it is spread over most of the constituents the results for minor constituents can be seriously in error. The cost of the mortars precludes large sizes so that the amount of sample which can be handled is relatively small, but is sufficient for the final particle size reduction of the actual analysis sample.

Tungsten Carbide

Tungsten carbide is another very hard substance and some workers have reported that even with hard materials contamination is normally less than 0·1%. The tungsten would not be expected to interfere with the analysis when present at this level; the bonding agent, usually cobalt, should be present in such small quantities as to be insignificant. More information on the degree of contamination to be expected when grinding very hard materials such as fused alumina is needed before the ultimate place of tungsten carbide in ceramic analysis can be determined. Information so far available suggests that tungsten carbide is liable to produce variable levels of contamination when used in some mechanical grinding equipment and should be used with care.

Boron Carbide

Boron carbide is extremely hard and introduces contamination about an order of magnitude lower than alumina. It appears suitable for most ordinary materials, the only important restriction being when boron itself is to be determined. It is relatively pure and the contamination is thus not likely to interfere with most analyses. Its chief disadvantage at present is its price (about £120) but it is possible that increasing demand will help to reduce this.

Mechanical

Various mechanical devices, which may be suitable for the final grinding stage, are available. Mechanical agate mortars (Fig. 3) have been on the market for a considerable time and can be used with confidence in place of hand grinding.

Fig. 3. Mechanical Agate Mortar. Reproduced by courtesy of Glen Creston Ltd.

Other machines include such devices as the Spex mixer mill (Fig. 4) and the Tema mill (Fig. 5), both of which offer several grinding media.

DRYING OVENS

Most ceramic materials are dried at 110°C before analysis and although accurate control of this temperature is not vital, too great a deviation from it is not desirable. The control of temperature is more important in the drying of precipitates, for instance any great increase in temperature over that specified when drying quinoline silicomolybdate is liable to result in some decomposition.

It is advisable, if practicable, to use electric ovens with thermostatic control, to ensure an even temperature and to avoid any contamination of the sample by fumes from gas heaters. Experience has shown that stainless steel is a better lining material than copper. Stainless steel appears to last indefinitely without noticeable corrosion when used at temperatures up to 150°C, whereas copper flakes badly even at 110°C.

Fig. 4. Spex Mixer Mill. Reproduced by courtesy of Glen Creston Ltd.

DESICCATORS

The particular pattern of desiccator chosen is a matter of personal preference. It is a mistake to use too small a desiccator as the heat dissipation from
the crucibles etc., appears to be slower. It is advisable to choose a size to
match the potential use. Silica gel can be conveniently used for most purposes
although it is by no means the most efficient of desiccants. A perforated zinc
disc should be placed over the desiccant but platinum ware should never be
allowed to rest directly on this. A desiccator plate with suitably sized holes to
take the platinum crucibles should be used. The size of hole should be arranged so that the crucible rests securely in it without its base coming into contact with the zinc disc below. Desiccator plates may be made of glazed or
unglazed porcelain.

BALANCES

The choice of suitable balances is important for any chemical laboratory.
For ceramic analysis a good quality four-figure analytical balance is essential
and also a rough balance for weighing out reagents to the second decimal
place, as it is most undesirable to use an accurate balance for this purpose.
It should be noted when considering the purchase of an analytical balance

Fig. 5. Tema Laboratory Disc Mill. Reproduced by courtesy of Tema (Machinery) Ltd.

that it is not likely to be replaced for a considerable length of time. Thus it is important to make a long-term assessment of the weighing requirements in the laboratory, admittedly a matter of some difficulty in small laboratories, particularly if recently established. The success of an analytical control scheme is almost certain to cause a large increase in the number of weighings. Apart from this and the time spent on weighing, there are technical considerations. Fast weighing not only saves time but gives increased accuracy since the material is on the balance for a shorter time, resulting in lower moisture adsorption. All these factors point to buying a fully automatic balance, preferably one of single-pan design. Admittedly the cost is approximately three times as much, but the old swinging-pointer balance is now an anachronism in an industrial ceramic laboratory. Its slow operation and the consequent danger of samples picking up moisture are so great that the newer type balance fully justifies the increased expenditure.

Almost all precision balance manufacturers now make a fully weight-loading single-pan balance. The Oertling, (Fig. 6), Stanton (Fig. 7), and Sartorius (Fig. 8) single pan balances all have partial release mechanisms to enable the smaller weights to be added without bringing the beam to rest.

The balance must be used carefully. Most balance manufacturers operate a maintenance service with skilled technicians and if the laboratory intends to

Fig. 6. Oertling Model R20 Balance. Reproduced by courtesy of L. Oertling Ltd.

carry out accurate analysis this scheme, if adopted, ensures that at least once or twice a year the balance is checked and when necessary brought back to peak performance. Little deterioration should occur between servicing if the balance is used correctly and with care.

The balance and case should be kept thoroughly clean and the pan(s) lightly brushed with a soft brush before and after use. The equilibrium of the balance should be checked before starting a weighing or series of weighings, that is, the balance with empty pan(s) should be adjusted to read zero on the scale or illuminated dial. It is also important to check that the balance is in a horizontal position, since errors here with the air-damped, fully weight-loading balances can be even more significant than with the swinging-pointer type. The object to be weighed, and weights—from 1 g upwards—should be put in position while the beam is at rest; with the swinging-pointer type **every weight loading** should be carried out with the beam at rest. The stated maximum load of the balance should never be exceeded, and all objects placed on the pan should be at room temperature. After weighing, all weights should be removed with the beam in the rest position. The balance case should be kept closed between weighings; it may be open during the rough weighing period associated with the gradual adding of sample, but when the final weight is being recorded the beam should be returned to rest, the case closed and the beam then released. No chemicals, powders or objects which might

Fig. 7. Stanton CL41 Balance. Reproduced by courtesy of
Stanton Instruments Ltd.

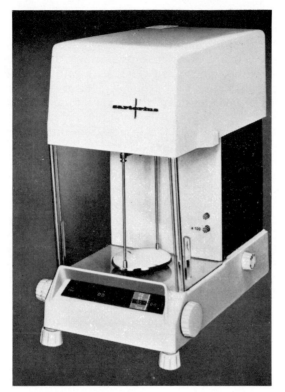

Fig. 8. Sartorius Single-pan Balance. Reproduced by courtesy of
the Scientific Instrument Centre Ltd.

damage the balance pan should be placed directly on it. Substances must
always be weighed into suitable containers, e.g. watch glasses and not paper.

PLATINUM WARE

Adequate stocks of platinum ware are essential and although such equipment is expensive it retains most of its value when it is returned as scrap and
should be regarded as a non-wasting asset. Platinum is extremely resistant to
chemical attack yet there are a number of circumstances in which it can suffer
damage, and there are certain chemicals which attack platinum readily. With
the appropriate care and attention platinum ware may be regarded as virtu-
ally indestructible. Almost all fusions carried out in platinum cause some
attack and consequent loss in weight. For this reason it is always desirable to
reduce the time and temperature of any fusion to a minimum, both these
factors being adjusted to the nature of the material being analysed. Sodium
peroxide sinters can be carried out safely in platinum within the temperature
range 480–490°C, i.e. below the melting point of this flux.

Platinum apparatus should always be heated in an oxidizing atmosphere otherwise there is a danger of a reaction with carbon. Also, when carrying out loss-on-ignition determinations or fusions, reduction of ferric oxide is liable to result in the formation of an alloy with iron, which on subsequent ignition re-oxidizes to the ferric state producing a blue coloration. To remove this, successive ignitions and treatments with hydrochloric acid are necessary, all of which are time-consuming. Thus it is desirable to complete either of the above processes in an electric furnace in which the atmosphere is normally sufficiently oxidizing to prevent any reduction.

Halogens, particularly chlorine (being most frequently encountered) readily attack platinum, so no oxidizing agent should ever be present when platinum is in contact with hydrochloric acid. Furthermore, platinum should never be allowed to come into contact with base metals such as lead, although low-temperature fusions of lead-bearing glazes can be carried out quite safely so long as no reduction takes place. In this case, fusion temperatures and times should be kept down to a minimum, consistent with the decomposition of the sample. Alkali hydroxides must not be fused in platinum.

Platinum should always be kept clean. In the normal course of an analysis all residues are removed as part of the respective determinations, but in other instances the residue should be removed as far as possible physically and an appropriate fusion carried out to remove last traces, even though these may be apparently invisible. Hot platinum should always be handled with platinum-tipped tongs, and care should be taken to see that only the tips come into contact with the platinum. These tongs should not be immersed in acid solutions, certainly not to a level where the unprotected base metal can come into contact with the liquid. In addition, it is not advisable to hold crucibles in the flame with such tongs for any length of time as this causes oxidation of the base metal underneath the tips with the risk of bursting them.

The life of platinum ware is considerably increased if its shape can be retained. Platinum is a very soft metal and any rough treatment will put creases into the ware. It is advisable to purchase, together with the platinum, a suitable plastic or boxwood former which should be utilized for reshaping the crucible or dish each time it has been used. To neglect this reshaping for even two or three uses can result in permanent deformation of the ware.

Experience has shown that a reasonable stock of platinum for one analyst engaged on ceramic analysis, particularly accurate analysis, would consist of one pair of large platinum dishes (3-in. dia.), one pair of small platinum dishes (2-in. dia.) for alkali determinations and two pairs of platinum crucibles of medium size. This assumes that the analyses will be carried out in duplicate.

BURNERS

The analyst will need about six gas points readily accessible to his bench. Two of these should be utilized for medium- or large-sized Méker burners, preferably with metal grids. Porcelain grids as normally supplied have a comparatively short life, and tend to burn back if cracks or hot spots develop. The remaining four points should be equipped with mushroom-headed burners (Fig. 9); these have the advantage that the heat is uniformly distributed over an area, allowing the liquid to be boiled in beakers without the danger of boiling over as a result of large bubbles forming at the hot spots. In addition, as the gas is issuing from a large number of small holes the degree of control is very high. These heads can be fitted on top of the usual Bunsen burner. It is useful to have an additional burner fitted with a pilot jet for ease of lighting the other burners.

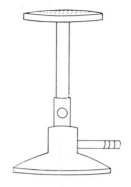

Fig. 9. Mushroom-headed burner.

Most fusions can be started safely over the mushroom burners and the flame is usually hot enough to carry through the fusion to the stage when it can be placed directly into the furnace. The flame is of a sufficiently oxidizing nature to prevent reduction of iron in the sample. The Méker burners should be brought into use to continue the fusion if and when the mushroom burners are not hot enough. Care should be taken to avoid using the Méker burners to their full heat if this is possible as there appears to be a tendency for reduction with the full flame. Platinum ware should not be allowed to rest with its base in the blue cones of any burner.

TRIANGLES

Triangles for supporting crucibles on a tripod are normally made of pipe-clay, silica or silica tubes threaded on wire. The last mentioned are the most generally used and have fewer disadvantages. If, after a time, the silica tends to flake or a fusion is spilt on to it the triangle should be discarded.

GAUZES

When boiling liquids in glass vessels, the most useful type of wire mesh gauze is that which has a circular mat of asbestos incorporated in the mesh. This helps to distribute the heat more evenly and also protects the wire from the direct action of the flame.

REAGENTS

The actual reagents on the bench will clearly depend on the range of work to be performed, but normally reagent bottles of the concentrated mineral acids and ammonia together with one or more dilutions of these and a limited range of other commonly used reagents are adequate. Dropping bottles, preferably of the T.K. type (Fig. 10) filled with reagents which may have to be added drop by drop in the course of an analysis, should also be available.

The reagents should be of analytical grade, except in a few isolated cases where this is unnecessary, e.g. hydrochloric acid for the Kipp's apparatus,

Fig. 10. T.K. type dropping bottle.

and where an analytical grade is not available. Other qualities of reagent should not be stocked in the laboratory as mistakes may arise in their usage. All reagents which are alkaline should be stored in polythene or similar plastic containers, as should solutions of EDTA, which are liable to extract the alkaline earths from the glass. Ammonia solution (concentrated) is often stored in an ordinary glass reagent bottle, but in this case it is advisable to allow the bottle to stand filled with the reagent for some time before use. The reagent which has been standing in the bottle is then discarded and the bottle taken into service. Prolonged storage of ammonia solution can result in some silica being dissolved from the glass.

WASH BOTTLES

One-litre flat-bottomed flasks are a convenient size for wash bottles and should be fitted with rubber (not cork) bungs bored for the mouthpiece and exit tube (Fig. 11). For hot liquids the neck of the flask should be bound, like a cricket bat, with one or two layers of string, thus forming an insulating layer and also a good grip.

Squeeze-type wash bottles (Fig. 12) made of plastic are useful for the final topping up to the graduation mark on volumetric flasks, and for washing precipitates where the wash liquid is objectionable, e.g. hydrogen sulphide wash liquors.

FURNACES

As previously stated, electric furnaces are preferable to gas; the atmosphere inside them is usually more oxidizing and they are more easily controlled, an ordinary energy regulator being adequate. Some of the furnaces used in the laboratories of the Association are wound with high-quality Nichrome wire with a nominal working temperature of up to 1150°C. One furnace can be run at a temperature of 1000°C for low-temperature ignitions and fusions, and another at 1200°C for high-temperature work. By keeping the furnaces at just over 700°C overnight and then raising the temperature next morning to that required, the life of the windings is from 9 to 12 months.

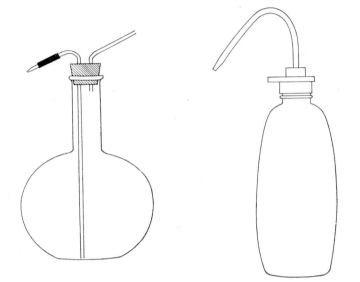

Fig. 11. Glass wash bottle.   Fig. 12. Plastic wash bottle.

Other furnaces are heated with silicon carbide bars operated by a transformer. These tend to have a longer life than wire-wound furnaces, particularly if they are automatically temperature controlled. In addition, replacing the heating elements is usually much simpler and faster as the wire often has to be "chipped out", having formed a slag at the point of failure.

DISTILLED WATER

Most commercial stills can be relied on to produce high-quality distilled water, such as is essential for analytical use, but it is preferable to choose one which does not incorporate too much copper. Small traces of copper often interfere with the detection of end-points in EDTA titrations. Electrically-heated stills are preferable to gas-heated ones because the latter can contaminate the water by fumes, particularly by traces of sulphurous gases.

De-ionized water is satisfactory for almost all the purposes for which distilled water is used. However, de-ionized water, especially towards the end of the life of the cartridge or resin bed, is liable to contain small amounts of silica which may become significant in a spectrophotometric determination of total silica.

STEAM BATHS

An efficient steam bath is essential, for example in silica dehydrations. If the single dehydration method is to be used exclusively then 3-in.-dia. holes will be adequate, but if analyses are performed by the classical method some 5-in.-dia. holes will be required to accommodate the larger porcelain basins. Electrically-heated steam baths are a distinct advantage, again allowing a better control and also having a greater safety factor, if left on overnight, thereby saving a large amount of time. Stainless-steel baths, if coated before use with an epoxy-resin paint, will last indefinitely, whereas copper and aluminium are prone to corrosion.

The rings used for covering the holes should be made of ceramic material rather than metal so as to avoid marking the platinum ware.

If the coagulation method is used it is preferable to use boiling water in beakers (with anti-bump granules) on the analyst's own bench. This reduces the risk of over-evaporation of the gel.

SAND BATHS

A sand bath provides a very convenient intermediate level of heat between a steam bath and a burner. The temperature applied to the base of a platinum dish or crucible can be varied either by moving it to different positions on the bath or by varying its depth in the sand. Electrical heating has advantages because of the control and the ability to leave the bath on overnight without the danger of fire.

HOT PLATES

An electrically-heated hot plate can be used for various purposes, particularly if it is fitted with an energy regulator control. The plate can be used for evaporating solutions at any speed, or for drying off alcohol from sodium hydroxide fusions since it provides a uniform heat over the whole of the surface in contact and so reduces the risk of bumping or spurting.

FUME CUPBOARDS

Efficient fume cupboards are a necessity since almost all complete analyses involve the evaporation of several millilitres of hydrofluoric acid. This can be a great hazard to health, and with the simultaneous use of perchloric acid another potential danger arises. Most silica dehydrations and other evaporations can be done satisfactorily in a glass-fronted fume cupboard lined with tiles or plastic. Asbestos should be avoided as this material contains all the constituents likely to be determined. Lead-lined fume cupboards have in the past been commonly used for carrying out evaporations of hydrofluoric acid, but these can now be replaced with plastic-lined cupboards. If a water wash by means of a spray is incorporated into the cupboard flue the danger from explosion with perchloric acid can be greatly reduced.

GLASSWARE

High-quality glassware of the high-resistance type should always be selected. Cheaper glassware is liable to serious attack with resulting contamination of the contents.

Of the various types of beaker on the market, the tall, straight-sided and conical forms are most popular. The latter are advantageous for boiling down and for certain titrations, but are inconvenient when a precipitate has to be "bobbied" out. For extended evaporations the squat form is recommended. All beakers should be lipped to facilitate pouring.

Except in a few instances titrations are normally carried out in conical flasks, the 500-ml size, and on occasion the 1-litre size, being most convenient.

Buchner flasks, say 600-ml and 1-litre, are needed for filtration under suction.

Separating funnels are needed for solvent extractions and from experience the 500-ml conical Squibb's type is the easiest to handle, though for some work 250-ml and 1-litre funnels should be used. The stoppers of all these should be made of glass as polythene is liable to attack by organic solvents.

Other essential glassware will include watch glasses of various sizes, glass rods of various lengths with the ends rounded off (in a gas flame) as stirrers, boiling tubes (see p. 32) and measuring cylinders from 10 ml up to 1 litre capacity, some of which can be stoppered to facilitate mixing *in situ* when making up solutions to approximate concentrations.

FILTRATION APPARATUS

It is important to select suitable funnels for filtration, and accurately moulded funnels (60°) with internal fluting or moulding, are essential. The funnel should be capable of holding a column of liquid when the paper is in position so as to utilize the pull of the column, thereby greatly increasing the speed of filtration. Many analysts, particularly in the metal industries, make use of paper pads as a replacement for a filter paper but this technique finds very little favour in ceramic analysis. The filter paper is normally folded in half and then in half again, after which it is opened out at the top to form the cone. A corner may be torn off the double thickness on the outside of the cone. The folded paper is bedded into the funnel and moistened with water. The column may now be formed by placing a finger over the outlet and filling the funnel with water. The water is then allowed to run through until the stem is full. Alternatively, by placing the flat of the hand over the filled funnel and pressing down, a piston action results in the formation of a column.

The choice of filter paper is dependent entirely on the work and as such is specified in the methods. The papers have been chosen so that all the precipitate should be retained and a clear filtrate obtained. There may be occasions when even the use of the finest paper will not separate all the solid; a double thickness of paper may help in this instance or it may be necessary to use other filtration techniques. Vacuum filtration is necessary on certain occasions.

The Buchner flask is used with a Buchner funnel, either with paper, or a sintered-glass mat (Fig. 13), and with filter candles such as are used for removing the clay during the determination of soluble salts.

Glass crucibles with sintered-glass mats are generally used for organic precipitates which can be dried at comparatively low temperatures in an air oven. Crucibles of grade-4 porosity are recommended as they give clear filtrates and are fast enough for all work, provided they are kept clean; the coarser grades are not sufficiently reliable. The most suitable size is approximately 40 mm tall and 40 mm diameter with a 30-mm-dia. mat. These should be cleaned before use by passing boiling chromic acid through in the reverse direction to normal filtration followed by boiling water until free from sulphates. After use, the crucibles are best cleaned by tapping out the bulk of the precipitate and then washing the mat with concentrated ammonia solution in the reverse direction, to avoid pulling the solution into the sintered mat. After washing with water the crucibles should be washed with hot chromic acid as detailed above. Sintered-glass funnels have proved to be advantageous for the filtration of calcium oxalate and lead sulphide precipitates; it is easier to dissolve the precipitate through and wash the mat than when a filter paper is used. On the other hand, the use of these funnels for the filtration of magnesium ammonium phosphate precipitates from the very alkaline mother liquor can result in some attack on the glass of the mat and the introduction of silica which may be co-precipitated during the second precipitation causing high results.

Another useful piece of equipment is Witt's apparatus (Fig. 14). To prevent breakage of the paper under pressure its tip is supported by a perforated platinum cone. The use of this apparatus helps to speed up slow filtrations

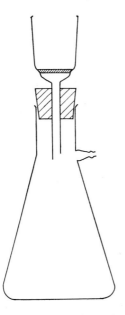

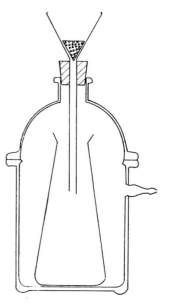

Fig. 13. Buchner flask, and funnel with sintered-glass mat.

Fig. 14. Witt's apparatus.

such as may occur after the dehydration of the silica in samples containing large amounts of zirconia or chrome and iron.

VOLUMETRIC GLASSWARE

Fairly large stocks of volumetric glassware are needed, particularly flasks ranging from 25 ml to 2 litres, although the intermediate sizes (100 to 500 ml) are most in demand. For work of normal accuracy grade B flasks are sufficiently accurate, but for work of the highest accuracy a limited quantity of grade A flasks may be required. It is worth noting, however, that the permissible error on a 250-ml grade B flask is O·2 ml, therefore the potential error when using the flask is only one part in 1250, i.e., on a 50% alumina content the error would not exceed 0·04%.

Volumetric flasks differ from pipettes and burettes in that they are calibrated to **contain** and not **deliver** a specified volume. This difference is generally accepted without thought but is nevertheless an important distinction. For accurate use the solution and washings should be transferred to the flask and then diluted with distilled water until the liquid is almost up to the graduation mark. After allowing the neck to drain for a few seconds the volume is carefully adjusted until the bottom of the meniscus is level with the mark. The stopper is then inserted tightly and the contents thoroughly mixed by inverting the flask a number of times.

Pipettes should be available in a range of sizes from 5 ml to 100 ml; the ordinary bulb pipette is most convenient and here again the choice of grade A or grade B will depend on the work to be carried out. A grade B 25-ml has a tolerance of 0·06 ml and on withdrawing an aliquot from a flask, e.g. for a spectrophotometric silica determination at the 75% level, the error may be up to 0·18%. This represents an appreciable part of the 0·5% tolerance for the method so that the actual pipettes used should be standardized, or grade A pipettes with about 50% lower tolerances stocked.

The points to be observed to ensure that the pipette actually delivers the specified volume are as follows.

(a) It should be rinsed twice with the particular solution, then filled to approximately 1 in. above the mark and the stem wiped with a piece of filter paper.

(b) With the pipette held vertically the excess of the solution should be slowly run out until the bottom of the meniscus coincides at eye-level with the graduation mark.

(c) Any drop adhering to the jet should be removed by touching the pipette against a glass surface and the pipette finally discharged into the appropriate vessel.

(d) The liquid should be allowed to run out freely; then after the specified additional drainage time the jet should be touched against the side of the vessel or in the case of a polythene vessel against the surface of the liquid. The small volume of liquid remaining in the jet must not be removed.

Various types of automatic pipettes are now available for the accurate addition of reagents, dilution and dispensing stock solutions. Possibly the most useful is the syringe type (Fig. 15) driven by an electric motor. Their main advantage is when set volumes of reagents, e.g. buffer and Solochrome cyanine solutions, have to be added repetitively in the course of a number of analyses.

The relative cheapness of automatic measures (Fig. 16) means that they are an economical proposition in the majority of laboratories. Sizes from 5 ml to

Fig. 15. Hook and Tucker "Auto-Spenser". Reproduced by courtesy of Hook and Tucker Ltd.

50 ml are very useful and help to cut down the amount of time spent on the addition of reagents in ordinary analysis, particularly spectrophotometric work.

A faster and more accurate type may well be preferred even if the cost is a little higher; this is the manual version of the syringe type dispenser (Fig. 17).

For most ceramic analyses two burette sizes are sufficient, 50-ml and the 10-ml semi-micro, the latter calibrated in 0·02 ml. The 50-ml is for general use, and the 10-ml is particularly useful for EDTA titrations of lime and magnesia, where the level of accuracy must be high to cater for the small amounts of constituents being determined. The choice of grade A or B depends on the requirements.

The burette should be rinsed at least twice with 3–4-ml portions of the solution to be used and the washings run out through the tap and jet. The filled burette is clamped vertically and the solution run out slowly until the bottom of the meniscus is level with the zero mark at eye-level, care being taken that no air bubbles remain in the jet, and that the last drop is removed

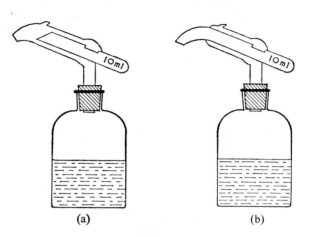

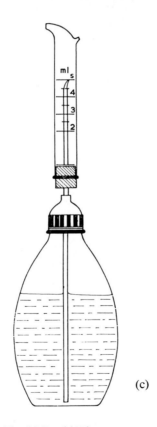

Fig. 16. Rapid Dispensers.

from the jet by touching against a glass surface. After titration, the volume used should be read by viewing the position of the bottom of the meniscus at eye-level.

Recently, various forms of piston burette (Fig. 18) have been introduced; these are advantageous where a laboratory has a number of repetitive titrations, for instance, EDTA titration for lime and magnesia, and alumina. By coupling a piston burette to a reservoir of the standard solution and using a magnetic stirrer (Fig. 18) much of the time and tedium can be eliminated from these volumetric procedures. Once the apparatus has been set up it is unnecessary to wash it out before and after each titration.

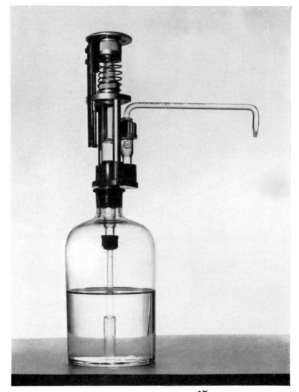

Fig. 17. Quickfit Dispenser. Reproduced by courtesy of J. A. Jobling and Co. Ltd.

Magnetic stirrers are invaluable for dissolving solids when making up standard or other solutions, and can also be used with a great saving of time for "beating" magnesium ammonium phosphate precipitates, particularly with very small amounts of magnesia.

Cleaning

Absolute cleanliness of volumetric glassware is essential; one frequent cause of error in the volume delivered by a pipette or burette is greasiness.

For normal working, a wash with "chromic acid" followed by flushing rapidly with water will remove any grease. A suitable solution for this purpose may be made as follows:

dissolve, by warming, about 5 g of potassium dichromate (commercial) in 1 litre of commercial sulphuric acid and allow to cool before use. Store in a glass-stoppered bottle. Sodium dichromate is more soluble in sulphuric acid (70 g/litre) and may be used if preferred.

The same mixture should be used regularly on ordinary glassware to keep it free from grease. During use, however, marks may develop on such glassware and stains on porcelain evaporating basins which chromic acid will not remove. The following cleaning mixture is effective in these cases: 5% HF, 33% HNO_3, 2% Teepol, 60% water. Use the cold solution. The mixture must

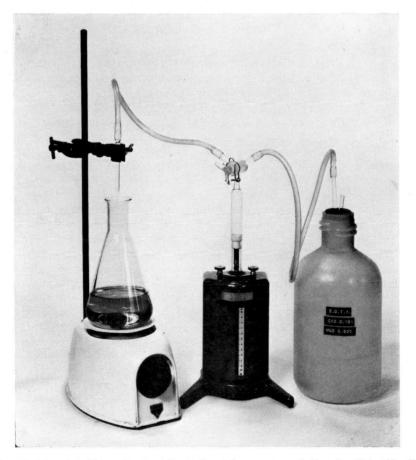

Fig. 18. Metrohm Piston Burette. Reproduced by courtesy of Shandon Scientific Co. Ltd. Bench Type Magnetic Stirrer. Reproduced by courtesy of Baird & Tatlock (London) Ltd.

not be used for cleaning volumetric glassware and, as it contains hydrofluoric acid, it should be handled with extreme care.

The beaker or basin should be rinsed out with the mixture, which in bad cases can be allowed to remain in contact for a few minutes, and then washed thoroughly with water.

NICKEL AND POLYTHENE BEAKERS

Nickel beakers are generally used for the aqueous extraction of sodium hydroxide fusions from crucibles, when the extract will be strongly alkaline. The use of glass beakers for this purpose can result in the solution becoming contaminated by constituents from the glass, particularly silica and alumina.

Polythene beakers are very useful for the preparation of alkaline reagents and EDTA solutions. They are also specified in the Rapid method for the determination of silica when the aliquot is to be made strongly alkaline.

FLAME PHOTOMETERS

If such equipment is to be used for the determination of soda, potash and lithia only, then it is advantageous to obtain a model with a low-temperature flame and using filters. The high-temperature instruments, although possessing a higher sensitivity, have many disadvantages. The spectra produced by an acetylene-air burner are appreciably more complicated requiring the use of a monochromator system to separate the lines of the various elements, thereby greatly increasing the cost of the equipment when compared with simple colour filters. In addition, interferences between one element or radical and another—hardly a problem at all with the low-temperature instrument— create many complications, as it is usually necessary either to remove the interfering elements or make up, for comparison, standard solutions which bear a close resemblance to the solution under analysis. In Britain, most laboratories engaged in ceramic analysis use an EEL instrument (Fig. 19). The design is simple and compact and the instrument has proved in practice to be very robust, rarely requiring attention outside the scope of the analyst. Using *coal gas*, if the acid concentration is reasonably controlled and the same technique is applied to making up the solutions, it has been shown that interferences, with one exception, are negligible. Large amounts of calcium can interfere in the determination of alkalis when the sample contains only small amounts of aluminium. Even this is readily overcome by adding aluminium to the sample solution; details of the actual technique will be found in Chapter 16.

pH METERS

Many methods now being developed are very dependent on pH control, either directly or by the addition of buffer solutions of accurately known pH. It is better to have the direct-reading instruments as these can be used for titrations such as may be encountered in the determination of borate.

BOILING TUBES

A Gernez boiling tube (Fig. 20) is an essential requisite for boiling and evaporating liquids, if bumping and mechanical loss are to be avoided. It is

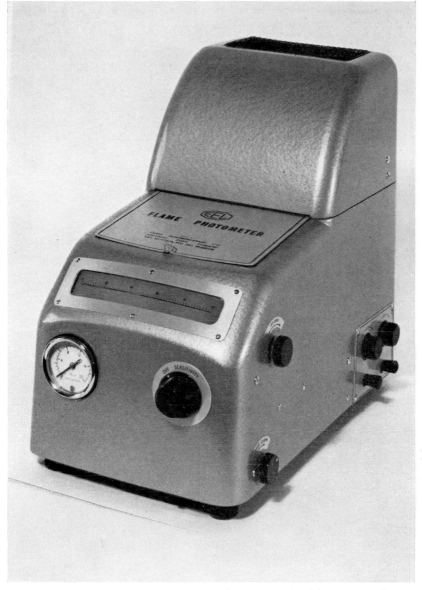

Fig. 19. EEL Flame Photometer Model 100. Reproduced by courtesy of
Evans Electroselenium Ltd.

made by heating a piece of medium-bore glass tubing, about 10 mm from the
end, until the walls soften and seal together. Alternatively, a piece of glass
tubing can be fused to the end of a piece of rod of similar size and nature,
the tube finally being cut off to a length of about 10 mm.

C

In use, the tube is placed with the end nearer the seal resting on the bottom of the containing vessel over the hottest part of the flame. It remains there during boiling. If a precipitation in the evaporated solution is desired, the vessel should be removed from the source of heat, the tube being then withdrawn and washed, including the hollow portion, and replaced by a solid rod. Before re-use, any water in the end is tapped out so as to provide the necessary air lock by which the device functions.

‹10mm›

Length to suit vessel

Fig. 20. Gernez boiling tube.

POLICEMEN OR BOBBIES

A policeman or "bobby" is a piece of rubber tubing sealed at one end. In use, it is stretched over the end of a glass rod, its function being to remove the last traces of a precipitate from a vessel. The type of policeman that should always be used is moulded in one piece from latex and is both pliable and durable. The policeman should be discarded immediately it shows any signs of disintegration.

OTHER ITEMS

Several other items are useful but not essential. Glass weighing scoops are very convenient for weighing out and introducing a solid into narrow-necked flasks. The tubing at the end of the scoop is inserted into the neck of the flask and the solid washed in with a jet of water from a wash bottle.

When determinations of borates or fluorides have to be carried out, a supply of glassware with standard cones and sockets is very desirable to facilitate quick setting up. The items kept in stock and the respective quantities must depend on the degree of usage.

Circular cork mats (about 5-in. dia.) not only protect the working bench from damage by hot vessels, but also prevent the pick-up of polish from the bench by the vessels.

Continuous use of calibration graphs, for the flame photometer and the spectrophotometer or absorptiometer, results in their deterioration. Such graphs are best stored between two sheets of plastic (e.g. Perspex) with the edges sealed by transparent plastic tape.

The Kipp's apparatus is a convenient way of generating gases such as carbon dioxide or hydrogen sulphide, particularly if either is in fairly regular use. The availability of these gases in cylinders means that they can be employed without all the tedium associated with cleansing the Kipp's apparatus and although the capital cost appears higher, it is doubtful if the real cost is, in fact, greater if analyst time is allowed for.

Part II. Methods of Analysis

CHAPTER 4

STOCK REAGENTS, INDICATORS AND STANDARD SOLUTIONS

INTRODUCTION

Unless otherwise stated, all reagents should be of analytical grade where available, and distilled water should be used throughout the analysis.

A list of reagents, as purchased from chemical suppliers, is given on p. 36 and is considered to represent the basic requirements of a laboratory actively engaged in ceramic analysis. Any reagents outside this list, required for a particular method, are noted as additional reagents at the beginning of the method.

Dilutions of acids and other liquid reagents are given as $(1+n)$, which indicates that one volume of the concentrated reagent should be added to n volumes of water and mixed. Care must be taken when adding concentrated sulphuric acid to water, as the solution becomes very hot and a sudden evolution of steam may cause spurting.

Solutions of solid reagents are given on a w/v basis; for instance barium chloride solution (100 g/litre) is made by dissolving 10 g of barium chloride in 100 ml of water. Solutions of solid reagents should be filtered if the solution is not clear.

All solutions of which the preparation does not conform to either of the above conventions are placed in the reagent section at the head of each method. Similar treatment is accorded to solutions which need to be freshly prepared or where there are other important factors e.g. the length of time the solution should be kept.

Reagent solutions which may attack glass should be stored in polythene bottles, e.g. ammonium molybdate, EDTA and alkaline solutions. As a precautionary measure against the possibility of base exchange, it is advisable to store standard solutions for flame photometry in polythene bottles.

The volume of a particular reagent to be made up must be judged from the amount used in each analysis and the frequency of use.

At the beginning of each method, as well as the special solutions mentioned above any additional stock reagents, indicator and standard solutions are listed.

It is hoped that this arrangement will make the method, as written, simple to use. There is a danger in over-coding reagents and apparatus, under the excuse of clarity, of making the volume of cross-reference so great as to make the method almost unintelligible.

STOCK REAGENTS

It is assumed that the following reagents will be stocked by a laboratory carrying out routine ceramic analysis. None of these reagents is included in the list preceding the individual methods.

Acetic acid: Glacial
Acetone
Aluminium: 99·99 %
Ammonia solution: $d = 0·88$
Ammonium carbonate
Ammonium chloride
Ammonium molybdate
Ammonium nitrate
Ammonium oxalate
Barium chloride
Boric acid
Bromine
Bromophenol blue
Calcein
Calcium carbonate
Chloroform: B.P. grade
Cupferron*
1:2-Diaminocyclohexane tetra-acetic acid (DCTA)
Diaminoethanetetra-acetic acid, di-sodium salt, dihydrate (EDTA)†
di-Ammonium hydrogen orthophosphate
Diethyl ether
2:4-Dinitrophenol
Dithizone
Ethanol (95 %): Industrial methylated spirit

Ethylene glycol bis-(2-aminoethyl)-tetra-acetic acid (EGTA)
Ferrous ammonium sulphate
Ferrous sulphide: Sticks
Fusion mixture‡
Hydrochloric acid: $d = 1·18$§
Hydrochloric acid: Commercial
Hydrofluoric acid: 40 % w/w
Hydrogen peroxide: (6 %)
Hydroxyammonium chloride
8-Hydroxyquinoline (oxine)
Iceland spar
Iron (wire)
Lithium sulphate
Magnesium: Turnings
Magnesium sulphate heptahydrate
Methanol
Methyl orange
Methyl red
Methylthymol blue complexone
Naphthol Green B
Nitric acid: $d = 1·42$
Oxalic acid
Perchloric acid: $d = 1·54$
1:10-Phenanthroline hydrate
Phenolphthalein
Phosphoric acid: $d = 1·75$
Polyethylene oxide‖

* Cupferron: The solid reagent should be stored in a tightly stoppered bottle in the presence of a piece or pieces of ammonium carbonate to prevent decomposition.

† Diaminoethanetetra-acetic acid, disodium salt (EDTA): Synonyms:
　　Complexone III
　　Disodium ethylenediaminetetra-acetate
　　Ethylenediaminetetra-acetic acid, disodium salt
　　Sequestric acid, disodium salt
　　Versenic acid, disodium salt.

‡ Fusion mixture: An equimolecular mixture of anhydrous potassium and sodium carbonates.

§ Hydrochloric acid $d=1·18$: In these methods concentrated hydrochloric acid of $d=1·18$ has been specified, but in laboratories in which an acid of $d=1·16$ is normally used, the volumes given in this book may be multiplied by the appropriate factor (1·14) if the difference would appear to be of significance.

‖ Union carbide 'Polyox' Resins WSR-35, WSR N-80, WSR-205, WSR N-750, or WSR N-3000 are suitable as sources of polyethylene oxide.

Potassium chloride
Potassium dichromate
Potassium hydroxide: Pellets
Potassium nitrate
Potassium periodate
Potassium permanganate
Potassium pyrosulphate
Potassium sulphate (anhydrous)
Potassium titanium oxalate
Silica¶
Silver nitrate
Sodium carbonate: Anhydrous
Sodium chloride

Sodium diethyl dithiocarbamate
Sodium dichromate: Commercial
Sodium hydroxide: Pellets
Sodium oxalate
Sodium peroxide
Sodium sulphate: Anhydrous
Solochrome Black 6B**
Stannous chloride
Sulphuric acid: $d = 1\cdot84$
Sulphuric acid: Pure
Thymolphthalein
Triethanolamine
Zinc: Pellets

INDICATORS

Alizarin red S (1 g/litre) Dissolve 0·1 g of alizarin red S in 100 ml of water-

Barium diphenylamine sulphonate (3 g/litre): Dissolve 0·3 g of barium diphenyl. amine sulphonate in 100 ml of warm water and cool.

Bromophenol blue (1 g/litre): Grind 0·1 g of bromophenol blue with 1·5 ml of sodium hydroxide solution (approx. 0·1N) and then dilute to 100 ml with water.

Calcein (screened): Mix, by grinding together, 0·2 g of Calcein, 0·12 g of thymolphthalein and 20 g of potassium chloride.

2:4-dinitrophenol (saturated solution): Dissolve 0·1 g of 2:4-dinitrophenol in 100 ml of hot water, allow to cool and filter.

Dithizone (0·25 g/litre): Dissolve 0·0125 g of dithizone in 50 ml of ethanol (95%). This solution will keep for about 1 week.

Methyl orange (0·5 g/litre): Dissolve 0·05 g of methyl orange in 100 ml of hot water, cool and filter.

Methyl red (1 g/litre): Dissolve 0·1 g of methyl red in 100 ml of ethanol (95%).

Methylthymol blue complexone: Mix, by grinding together, 0·2 g of methyl-thymol blue complexone and 20 g of potassium nitrate.

p-Nitrophenol (10 g/litre): Dissolve 1 g of p-nitrophenol in 70 ml of ethanol (95%) and dilute to 100 ml.

Phenolphthalein (10 g/litre): Dissolve 1 g of phenolphthalein in 100 ml of ethanol (95%).

Phenolphthalein-thymol blue mixed indicator: Dissolve 0·04 g of phenolphtha-lein in 40 ml of ethanol (95%) and 0·06 g of thymol blue in 60 ml of ethanol (95%). Mix the two solutions.

¶ Silica: A quartz or sand of high purity: the actual purity to be determined and the calibration graph adjusted accordingly.
   ** Also known as Eriochrome Blue-black B.

Solochrome Black 6B (also known as Solochrome Blue-black B): Mix, by grinding together, 0·5 g of Solochrome Black 6B and 20 g of sodium chloride.

Starch (10 g/litre): Mix 1 g of soluble starch into a thin paste with 10 ml of water. Pour, with stirring, into 90 ml of boiling water. Boil for 1 min and cool. This solution will keep for about 1 week.

Thymol blue (0·4 g/litre): Dissolve 0·04 g of thymol blue in 20 ml of ethanol (95%). Add 0·9 ml of sodium hydroxide solution (approx. 0·1N) and dilute to 100 ml with water.

STANDARD SOLUTIONS

NOTE: Some solutions may, on keeping, show signs of mould formation; this applies particularly to alkali and aluminium solutions. If this occurs the solution must be discarded.

Aluminium (1·0 mg Al_2O_3/ml): Wash the metal with diluted hydrochloric acid to remove any oxide film, then with water and finally with alcohol followed by ether. Weigh 0·5293 g of aluminium metal (99·99% Al) into a nickel beaker provided with a cover and dissolve in 50 ml of sodium hydroxide solution (40 g/litre). Transfer the solution to a 100-ml beaker containing 15 ml of hydrochloric acid (d=1·18), adding the rinsings from the nickel beaker. Cool and dilute to 1 litre in a volumetric flask.

Calcium (0·05M): Dissolve 5·0045 g of dried (150°C) calcium carbonate in a slight excess of diluted hydrochloric acid (1+4), boil to expel carbon dioxide, cool and dilute to 1 litre in a volumetric flask. 1 ml of this solution ≡2·804 mg CaO.

Calcium (1·0 mg CaO/ml): Dissolve 1·785 g of dried (150°C) calcium carbonate in a slight excess of diluted hydrochloric acid (1+4), boil to expel carbon dioxide, cool and dilute to 1 litre in a volumetric flask.

Chromium (1·0 mg Cr_2O_3/ml): Dissolve 1·9354 g of recrystallized, dried (150°C) potassium dichromate in water and dilute to 1 litre in a volumetric flask.

1:2-diaminocyclohexane tetra-acetic acid (*DCTA*) (0·05M approx): Dissolve 18·2175 g of DCTA in 500 ml of water by the progressive addition of the minimum amount of potassium hydroxide solution (250 g/l). (Approximately 25 ml should be required.) Dilute to 1 litre in a volumetric flask. Store in a polythene bottle. Standardize against both standard magnesium and standard zinc solutions as follows:

> *Against magnesium.* Transfer 50·0 ml of the standard magnesium solution (5 mg MgO/ml) to a 500-ml conical flask. Add 100·0 ml of the DCTA solution. Then add 2 g of ammonium chloride and 25 ml of ammonia solution (d=0·88). Titrate with the DCTA solution, using Solochrome Black 6B as indicator, from red through purple to the last change to a clear ice blue. (1 ml of 0·05M DCTA≡2·016 mg MgO).

> *Against zinc.* Transfer 50·0 ml of the DCTA solution to a 500-ml conical flask and add 5–6 drops of hydrochloric acid (d=1·18). Add a few drops of bromophenol blue indicator solution and then add ammonium acetate

buffer solution (see page 45) until the indicator turns blue, follow by 10 ml in excess.

Add a volume of ethanol (95%) equal to the total volume of the solution, follow by 1–2 ml of dithizone indicator solution and titrate with the standard zinc solution (0·05M) from blue-green to the first appearance of a permanent pink colour. (1 ml of 0·05M DCTA $\equiv$2·55 mg Al_2O_3).

EDTA (0·05M): Dissolve 18·6125 g of the salt in warm water, filter, cool and dilute to 1 litre in a volumetric flask. Store in a polythene bottle.

For use in the determination of alumina, standardize against standard zinc solution (0·05M) as follows:

Pipette 50·0 ml of the EDTA solution into a 500-ml conical flask and add 5–6 drops of hydrochloric acid (d=1·18). Add a few drops of bromophenol blue indicator solution and then add ammonium acetate buffer solution (see page 45) until the indicator turns blue followed by 10 ml in excess. Add a volume of ethanol (95%) equal to the total volume of the solution followed by 1–2 ml of dithizone indicator and titrate with standard zinc solution (0·05M) from blue-green to the first appearance of a permanent pink colour. (1 ml of 0·05M EDTA $\equiv$2·55 mg Al_2O_3).

EDTA (5 g/litre): Dissolve 5 g of the salt in warm water, filter, cool and dilute to 1 litre. Store in a polythene bottle.

Standardize against standard calcium and magnesium solutions as follows:

Calcium: Pipette 25·0 ml of standard calcium solution (1·0 mg CaO/ml) into a 500-ml conical flask, add 10 ml of potassium hydroxide solution (250 g/litre) and dilute to about 200 ml. Add about 0·015 g of screened Calcein indicator and titrate with the EDTA solution (5 g/litre) from fluorescent green to pink.

Magnesium: Pipette 25·0 ml of standard magnesium solution (1·0 mg MgO/ml) into a 500-ml conical flask, add 20 drops of hydrochloric acid (d=1·18), 20 ml of ammonia solution (d=0·88) and dilute to about 200 ml. Add about 0·04 g of methylthymol blue complexone indicator and titrate with the EDTA solution (5 g/litre) from blue to colourless.

If the EDTA solution is to be used for the determination of lime and magnesia in frits or glazes in the presence of BAL, Solochrome Black 6B indicator should be used for the standardization against standard magnesium solution. Add about 0·07 g of the indicator and titrate from wine red to the last change to a clear blue.

Ethylene glycol bis-(2-aminoethyl)-tetra-acetic acid (EGTA) (0·05M approx). Dissolve 19·0174 g of EGTA in 500 ml of water by the progressive addition of the minimum amount of potassium hydroxide solution (250 g/litre). (Approx. 25 ml should be required.) Dilute to 1 litre in a volumetric flask. Store in a polythene bottle. Standardize against the standard calcium solution as follows:

Transfer 50 ml of magnesium sulphate heptahydrate solution (6 g/litre) to a 250-ml volumetric flask and add 50 ml of the EGTA solution. Dilute to 150 ml, add potassium hydroxide solution (250 g/litre) until no further precipitation occurs and then add 10 ml in excess, followed by 10 ml of Magflok solution (20 g/litre). Dilute to 250 ml, shake and allow to stand for about 10 min to settle. Filter through a dry coarse filter paper (125 mm) into a dry beaker. Pipette 200 ml of the filtrate into a 500-ml conical flask and add 15 ml of potassium hydroxide solution (250 g/litre). Titrate with standard lime solution (0·05M) using screened Calcein as indicator to the first appearance of green fluorescence. (1 ml of 0·05M EGTA $\equiv$ 2·804 mg CaO).

Ferrous ammonium sulphate (approx. 0·1N): Dissolve 39·2 g of ferrous ammonium sulphate in 1 litre of diluted sulphuric acid (1 + 4). Standardize against standard potassium dichromate solution (0·1N).

Fluoride (1·0 mg F/ml): Dissolve 1·1050 g of sodium fluoride in water and dilute to 500 ml in a volumetric flask. Store in a polythene bottle.

Hydrochloric acid (approx. 2N)—used in the British Ceramic Research Association method for lead solubility: Dilute 175 ml of hydrochloric acid ($d = 1·18$) to 1 litre. Standardize gravimetrically against Iceland spar or volumetrically against sodium carbonate using methyl orange as indicator. [Let x be the normality of the acid; then 68·5/2x ml of this solution will give 500 ml of hydrochloric acid (0·25% HCl w/w).]

Hydrochloric acid (0·25% HCl w/w)—used in the Government method for lead solubility: Dilute 175 ml of hydrochloric acid ($d = 1·18$) to 1 litre. (This acid will be approx. 2N) Standardise gravimetrically against Iceland spar, or volumetrically against sodium carbonate using methyl orange as indicator. Let x be the normality of the acid. Using a burette, transfer 68·5/x ml of the standardized acid into a 1-litre volumetric flask and dilute to the mark.

Hydrochloric acid (approx. 0·5N): Dilute 45 ml of hydrochloric acid ($d = 1·18$) to 1 litre. Standardize gravimetrically against Iceland spar, or volumetrically against sodium carbonate using methyl orange as indicator.

Hydrochloric acid (approx. 0·1N): Dilute 10 ml of hydrochloric acid ($d = 1·18$) to 1 litre. Standardize gravimetrically against Iceland spar, or volumetrically against sodium carbonate using methyl orange as indicator.

Iron (0·1 mg Fe_2O_3/ml): Dissolve 0·4911 g of ferrous ammonium sulphate in water and add 10 ml of diluted sulphuric acid (1 + 1) and 5 ml of hydrogen peroxide (6%). Boil for 15 min to decompose the excess of hydrogen peroxide. Cool and dilute to 1 litre in a volumetric flask.

Lithium—concentrated standard solution A ($Li_2O \equiv 400$ ppm): Dehydrate lithium sulphate monohydrate by drying at 150°C for 24 h. Dissolve 1·4719 g of the anhydrous salt in water and dilute to 1 litre in a volumetric flask. Store in a polythene bottle.

Lithium standard solution B ($Li_2O \equiv 40$ ppm): Dilute 50·0 ml of the lithium concentrated standard solution A to 500 ml in a volumetric flask and mix.

Magnesium (1·0 mg MgO/ml): Wash the metal in diluted hydrochloric acid to remove any oxide film on the turnings, then with water and finally with alcohol followed by ether. Dissolve 0·6032 g of the oxide-free metal in a slight excess of diluted hydrochloric acid (1 + 9), cool and dilute to 1 litre in a volumetric flask.

Magnesium (5 mg MgO/ml): Wash the metal in hydrochloric acid (1 + 9) to remove any oxide film on the turnings, then with water and finally with alcohol followed by ether. Dissolve 3·016 g of the oxide-free metal in a slight excess of hydrochloric acid (1 + 9), cool, dilute to 1 litre in a volumetric flask and mix.

Manganese (0·1 mg MnO/ml): Dilute the calculated volume of ready standardized (approx. 0·1N) potassium permanganate to 1 litre in a volumetric flask; with exactly 0·1N potassium permanganate 70·5 ml are required.

Nitric acid (approx 1N): Add 63 ml of nitric acid ($d = 1·42$) to 500 ml of water, cool and dilute to 1 litre. Standardize gravimetrically against Iceland spar, or volumetrically against sodium carbonate using methyl orange as indicator.

Phosphate (0·1 mg P_2O_5/ml): Dissolve 0·1917 g of potassium dihydrogen orthophosphate in water and dilute to 1 litre in a volumetric flask. Store in a polythene bottle.

Potassium dichromate (0·1N): Dissolve 4·9032 g of the recrystallized, dried (150°C) salt in water and dilute to 1 litre in a volumetric flask.

Potassium dichromate (0·05N): Dissolve 2·4516 g of the recrystallized, dried (150°C) salt in water and dilute to 1 litre in a volumetric flask.

Potassium permanganate (approx. 0·1N): Dissolve about 3·2 g of potassium permanganate in 1 litre of water to obtain an approx. 0·1N solution; boil for 5 min, filter through a crucible with a sintered-glass mat (porosity 4) and cool. Standardise against sodium oxalate, and store in an amber glass bottle.

Potassium/Sodium—concentrated standard solution A ($K_2O \equiv 400$ ppm; $Na_2O \equiv 100$ ppm): Dissolve 0·7400 g of anhydrous potassium sulphate (dried at 150°C) and 0·2292 g of anhydrous sodium sulphate (dried at 150°C) in water and dilute to 1 litre in a volumetric flask. Store in a polythene bottle.

Potassium/sodium standard solution B ($K_2O \equiv 40$ ppm, $Na_2O \equiv 10$ ppm): Dilute 50·0 ml of the potassium/sodium concentrated standard solution A to 500 ml in a volumetric flask and mix.

Silica (approx. 0·5 mg SiO_2/ml): Fuse 0·5000 g of pure silica (>99% purity, see page 37) in 5 g of anhydrous sodium carbonate in a platinum crucible. Cool and dissolve in water in a polythene beaker. Cool and dilute to 1 litre in a volumetric flask. Store in a polythene bottle. (The actual purity of the silica must be determined and the calibration graph adjusted accordingly.)

Sodium hydroxide (approx. 1N): Dissolve 40 g of sodium hydroxide pellets in 500 ml of freshly boiled, cooled, distilled water, cool and dilute to 1 litre. Store in a polythene bottle. Standardize according to the method for which the solution is required, either against standard hydrochloric acid (approx. 0·5N), using phenolphthalein-thymol blue mixed indicator (small amounts of phosphate) or against standard nitric acid (approx. 1N) using phenolphthalein as indicator (phosphate in bone ash).

Sodium hydroxide (approx. 0·1N): Dissolve 4 g of sodium hydroxide pellets in 250 ml of freshly boiled, cooled, distilled water, cool and dilute to 1 litre. Store in a polythene bottle. Standardize against standard hydrochloric acid (approx. 0·1N), using phenolphthalein as indicator.

Thorium nitrate (approx. 0·1N): Dissolve 14·7045 g of thorium nitrate, $Th(NO_3)_4 . 6H_2O$, in water and dilute to 1 litre. Standardize against standard fluoride solution (1·0 mg F/ml) as follows:

> Pipette 10·0 ml of standard flouride solution (1·0 mg F/ml) into a 500-ml conical flask marked to indicate a volume of 90 ml. Dilute to 90 ml. Add 1 drop of phenolphthalein indicator and sodium hydroxide solution (approx. 0·1N), drop by drop, until a permanent pink colour is obtained.
>
> Add 10 ml of starch indicator and diluted perchloric acid (1 + 9), drop by drop, until the colour is discharged. Add 1 drop in excess and then add 0·5 ml of chloracetic acid buffer solution (see page 214).
>
> Add 5 drops of alizarin red S indicator and titrate with the thorium nitrate solution (approx. 0·1N) until the first appearance of a faint pinkish tint.

Titanium (1·0 mg TiO_2/ml): Ignite pure TiO_2 and then fuse 1·000 g with 10 g of potassium pyrosulphate. Allow to cool and dissolve, at a low temperature to avoid hydrolysis, in 200 ml of water to which 20 ml of sulphuric acid ($d = 1·84$) has been cautiously added. Cool, dilute to 1 litre in a volumetric flask and mix.

The following may be used as an alternative method. Evaporate potassium titanium oxalate almost to dryness with sulphuric acid ($d = 1·84$). Extract the residue with water and boil to hydrolyse the titanium sulphate. Filter, wash the residue with water until it is free from sulphates and then ignite to TiO_2. Fuse 1 g of the pure titanium dioxide with 10 g of potassium pyrosulphate, allow to cool and dissolve, at a low temperature to avoid hydrolysis, in 200 ml of water to which 20 ml of sulphuric acid ($d = 1·84$) has been cautiously added. Cool, dilute to 1 litre in a volumetric flask and mix.

Titanium (0·64 mg TiO_2/ml)—used in routine control analysis: Transfer 2·8319 g of potassium titanium oxalate to a conical flask. Add 100 ml of sulphuric acid ($d = 1·84$), gradually heat to boiling and boil for 5–10 min. Cool, dilute to 300 ml with water, taking great care to avoid spurting, and cool again. Dilute to 1 litre in a volumetric flask.

Zinc (0·05M): Wash the metal in diluted hydrochloric acid to remove any oxide film on the pellets, then with water and finally with alcohol followed by ether. Dissolve 3·2685 g of the oxide-free metal in 10 ml of hydrochloric acid ($d = 1·18$) and about 50 ml of water. Cool and dilute to 1 litre in a volumetric flask. 1 ml of this solution $\equiv 2·55$ mg Al_2O_3.

CHAPTER 5

ANALYSIS OF HIGH-SILICA MATERIALS, ALUMINOSILICATES AND ALUMINOUS MATERIALS:

A. Coagulation Method

GENERAL

The British Standard Methods (B.S. 1902: Parts 2A and 2B) are being amended to allow the coagulation method for silica to be used as an alternative to the single dehydration (accurate) method (B.S. 1902: Part 2D, 1969). The only alterations are in the procedures for decomposing the sample and for insolubilizing the silica.

As the coagulation method is now preferred, it is included in the full method; the experimental details for the single dehydration method are added at the end of the chapter. The advantages of coagulation are that the silica residue is less, and the method is appreciably faster without loss of any accuracy.

PRINCIPLE OF THE METHOD

The loss on ignition is determined by heating the sample to 1000°C for 30 min. The same sample, or a freshly weighed portion, is fused in fusion mixture/boric acid at 1200°C in a 70-mm-dia. platinum dish for from 5 up to 30 min, depending on the alumina content; the melt is then dissolved in hydrochloric acid with a little sulphuric acid to assist in preventing hydrolysis of titanium salts. The solution is evaporated for a few minutes to form a stiff gel (in the case of samples containing less than about 20% SiO_2, which do not usually gel, the evaporation is continued for 30 min), and the silica coagulated with polyethylene oxide. The precipitate is filtered and washed, first with diluted hydrochloric acid and then hot water, transferred to a platinum crucible and ignited and weighed before and after hydrofluoric acid treatment. The difference in weights gives the gravimetric silica figure.

The residue after hydrofluoric acid treatment is fused in sodium carbonate/boric acid or potassium pyrosulphate; the solution of the melt is added to the filtrate from the silica, and the combined solution made up to volume.

Aliquots of the stock solution are used for the spectrophotometric determination of ferric oxide utilizing 1:10 phenanthroline, after reduction with hydroxyammonium chloride, of titania using hydrogen peroxide, with phosphoric acid to prevent the interference of iron, and of residual silica in solution by either the yellow colour using ammonium molybdate or the blue colour after the reduction of the silicomolybdate. In most cases, when the phosphate is not significant, the ammonium molybdate procedure is simpler and adequate, but in the presence of phosphate or on the rare occasion when the solution tends to become cloudy after the addition of the molybdate, the molybdenum blue procedure is used.

Other aliquots may be used for spectrophotometric determinations of phosphate by the addition of ammonium molybdate and ammonium vanadate, having first removed chlorides by nitric acid evaporation; and of manganese by the oxidation to permanganate by potassium periodate, first removing chlorides by nitric and sulphuric acid evaporation.

A further aliquot is used for the volumetric determination of alumina. After increasing the acidity with hydrochloric and sulphuric acids, the iron and titanium are precipitated with cupferron and the precipitate and excess reagent removed by solvent extraction with chloroform. The solution is then neutralized with ammonia, excess EDTA is added and the mixture buffered. The solution is now boiled to ensure complexing of all the aluminium, cooled and the excess EDTA back-titrated with zinc solution using dithizone as indicator in the presence of alcohol.

The method for the determination of alumina in which EDTA is replaced by DCTA may be used as an alternative to the above. This has some advantages, one of which is greater speed, and is of equal accuracy. The procedure is described in detail in Chapter 17.

For aluminous materials the lime and magnesia determinations are also made on the stock solution, by the method described below.

Alkalis are determined after hydrofluoric acid treatment of the sample; for high-silica and aluminosilicate materials, nitric and sulphuric acids are used and usually effect complete dissolution, but for aluminous materials filtration is necessary. The soluble portion of the residue is dissolved in nitric acid (filtered if necessary) and made up to volume.

A portion of the solution is used for the flame-photometric determination of the alkalis and two further aliquots are withdrawn for the determination of lime and of lime plus magnesia. In both titrations triethanolamine is used to complex the ammonia-group oxides. For the determination of lime, the solution is made alkaline with potassium hydroxide and titrated with EDTA using screened Calcein as indicator. For the sum of lime and magnesia, ammonia is used to make the solution alkaline before titrating with EDTA with methylthymol blue complexone as indicator.

Failure to achieve analytical totals may necessitate further determinations, sulphur being the most usual, but in certain materials fluorine, zirconium and barium are not uncommon.

Routine Control Application

The coagulation method lends itself readily to routine control analysis. By a simple modification, a full analysis can be completed in 1 day, with only minimal loss of accuracy. Experience has shown that, provided satisfactory fusion has been achieved, as is usually the case, residues after hydrofluoric acid treatment are less than 2 mg ($= 0.2\%$). Only about 0.5 mg of this is Al_2O_3, with lesser amounts of the other elements sought (alkali sulphate usually accounts for about 1 mg). If this residue is ignored, the magnitude of the errors are very small.

Thus it is possible to carry out the remaining determinations on the silica filtrate, while the silica precipitate is being treated to provide the "gravimetric" silica figure.

REAGENTS

Additional Reagents

Ammonium vanadate.

Prepared Reagents

Ammonium acetate (approx. 40%): Dilute 570 ml of acetic acid (glacial) to 1700 ml with water and add carefully 570 ml of ammonia solution ($d = 0.88$). Mix, cool and adjust to pH 6·4–6·8, using either acetic acid or ammonia solution.

Ammonium acetate (approx. 10%): Dilute the ammonium acetate solution (approx. 40%) with three times its volume of water.

Ammonium acetate buffer: Add 120 ml of acetic acid (glacial) to 500 ml of water followed by 74 ml of ammonia solution ($d = 0.88$). Mix, cool and dilute to 1 litre.

Ammonium molybdate (80 g/litre): Dissolve 80 g of ammonium molybdate in 1 litre of water; filter if necessary. Store in a polythene bottle and discard after 4 weeks or earlier if any appreciable deposit of molybdic acid is observed.

Cupferron (60 g/litre): Dissolve 3 g of cupferron in 50 ml of cold water and filter. This solution must be freshly prepared.

If the reagent is discoloured or gives a strongly coloured solution a new stock should be obtained. The solid reagent should be stored in a tightly stoppered bottle in the presence of a piece of ammonium carbonate to prevent decomposition.

8-Hydroxyquinoline (50 g/litre): Dissolve, by warming, 50 g of 8-hydroxy-quinoline in 120 ml of acetic acid (glacial). Dilute to about 700 ml, filter, cool and dilute to 1 litre.

1:10 Phenanthroline (10 g/litre): Prepare enough solution for immediate use at a concentration of 0·1 g of 1:10 phenanthroline hydrate in 10 ml of diluted acetic acid (1 + 1).

Polyethylene oxide (2·5 g/litre): Add 0·5 g of polyethylene oxide* to 200 ml of water slowly with stirring, preferably on a mechanical stirrer, until dissolved. Discard after 2 weeks.

Silver nitrate (1 g/litre)—used for testing for chlorides: Dissolve 0·1 g of silver nitrate in 100 ml of diluted nitric acid (1 + 9).

Stannous chloride (10 g/litre): Dissolve, by warming, 1 g of stannous chloride in 1·5 ml of hydrochloric acid ($d = 1.18$). Cool, dilute to 100 ml and mix. This solution will not keep longer than 24 h.

Sulphuric–nitric acid mixture: To 650 ml of water, add 100 ml of diluted sulphuric acid (1 + 1) and 250 ml of nitric acid ($d = 1.42$).

*Union Carbide "Polyox" resins WSR 35, WSR-N-80, WSR 205, WSR-N-750 or WSR-N-3000 are suitable as a source of polyethylene oxide.

Indicators

Bromophenol blue
Calcein (screened)
2:4 Dinitrophenol
Dithizone
Methyl red
Methylthymol blue complexone

Standard Solutions

Aluminium (1·0 mg Al_2O_3/ml)
EDTA (0·05M)
EDTA (5 g/litre)
Zinc (0·05M)

PREPARATION OF SAMPLE

The sample for analysis should be ground to pass completely through a 120-mesh B.S. test sieve. A non-metallic (e.g. nylon bolting-cloth) sieve is preferable.

High-silica materials and aluminosilicates containing up to 45% Al_2O_3 may be ground in agate, or in iron with subsequent treatment with a magnet.

Aluminous materials may be ground in alumina or in iron mortars; in the latter case the ground sample should be treated with a magnet. It may be necessary to prepare a sample by each method and carry out the main analysis on the iron-ground sample with a ferric oxide determination on the alumina-ground sample and then calculate the results to the latter iron content. Any metallic iron introduced will be oxidized to the ferric state by the heat treatment during the determination of the loss on ignition. Correction must be made for the difference in ferric oxide content of the two samples, expressed as metallic iron, the state in which it is present when weighing out the sample.

The use of a boron carbide mortar permits the grinding of aluminous materials with the minimum of contamination, thus obviating the necessity for preparing two separate portions of the sample.

PROCEDURE

Determination of Loss on Ignition

Weigh 1·000 g of the finely ground, dried (110°C) sample into a platinum crucible. Place the crucible in a muffle furnace and slowly raise the temperature to 1000 ± 25°C. Ignite to constant weight at this temperature; 30 min ignition is usually sufficient.

For routine work, it is possible to start the ignition over a low mushroom flame, slowly increasing the temperature to full heat over a period of about 20 min, after which the crucible is transferred to a furnace at 1000°C for 30 min.

Determination of Silica, Alumina, Ferric Oxide, Titania, Manganese Oxide and Phosphorus Pentoxide

Decomposition of the Sample

NOTE: For routine work, the loss on ignition may be carried out in the platinum dish and the same portion of sample used for the fusion. This technique is only admissible if the amount of sintering of the sample after the determination of loss on ignition does not prevent adequate mixing of the sample and flux.

Weigh 1·000 g of the finely ground, dried (110°C) sample into a platinum dish (70-mm dia., 40-mm deep, of effective capacity 75 ml). *For high-silica and aluminosilicate materials* add 3 g of fusion mixture and 0·4 g of boric acid, *for aluminous materials* add 1·5 g of fusion mixture and 0·2 g of boric acid and mix well to form a charge of about 50-mm dia. in the centre of the dish. Cover with a lid.

Heat over a mushroom flame, cautiously at first, then gradually raise the temperature to the full heat of the burner over a period of about 10 min. In some cases it may be necessary to transfer the dish to a Méker burner to melt the flux effectively.

For high-silica and aluminosilicate materials heat the dish in a muffle furnace for 10 min at 1200°C, but for *aluminous materials* the time of fusion will need to be varied from 15 to 30 min, the greater time being necessary for higher alumina contents. Remove the dish from the furnace and allow to cool.

If the fusion appears blue or blue-green, indicating the presence of manganese, add 2 ml of ethanol (95%) to the contents of the dish prior to dissolution.

Add 15 ml of hydrochloric acid ($d = 1·18$), 1 ml of diluted sulphuric acid ($1 + 1$), 10 ml of water and transfer the dish to a steam bath to facilitate dissolution of the melt. If the solution gels before dissolution is complete, it may be necessary to stir the gel gently with a glass rod. When solution is complete, remove the lid and wash any spray adhering to it into the dish with the minimum amount of water.

Determination of Gravimetric Silica

Allow the dish to remain on the steam-bath until a stiff gel is formed. For samples containing less than about 20% of silica, gelling may not occur; in this case allow the dish to remain on the steam-bath for 30 min. Loss of hydrochloric acid will occur during prolonged evaporation; it is desirable to add an appropriate amount of diluted hydrochloric acid ($1 + 1$) to replace this loss in the latter instance. When a stiff gel has formed, or after 30-min evaporation, thoroughly mix a Whatman accelerator tablet into the gel until the mixture is homogeneous, ensuring that the gel adhering to the sides of the dish is also broken up. Add slowly, with stirring, 5 ml of polyethylene oxide coagulant solution (2·5 g/litre) and mix well, ensuring that any gel adhering to the sides of the dish is brought into contact with the coagulant. Add 10 ml of water, mix and allow to stand for 5 min.

Filter through a 110-mm No. 42 Whatman paper, transferring the silica to the filter with hot, diluted hydrochloric acid (1 + 19), scrubbing the dish with a rubber-tipped glass rod. Wash the precipitate five times with hot, diluted hydrochloric acid (1 + 19) and then with hot water until free from chlorides (usually to a volume of about 400 ml). The silica precipitate obtained by this method is more voluminous than that obtained by dehydration and so care must be taken to ensure that the mass of precipitate is broken up thoroughly during the washing. Reserve the filtrate and washings.

Transfer the paper and precipitate to an ignited and weighed platinum crucible. Ignite at a low temperature until the precipitate is free from carbonaceous matter and then heat in a muffle furnace at 1200°C to constant weight, 30 min usually being sufficient. (The ignition should be started at a lower temperature than that used in the standard method, or a noticeable amount of carbon may remain after ignition. In most cases a trace of carbon remains but its weight appears to be negligible and does not affect the results.)

Moisten the contents of the cold crucible with water, add 5 drops of diluted sulphuric acid (1 + 1) and 10 ml of hydrofluoric acid (40% w/w). Evaporate to dryness on a sand bath in a fume cupboard. For the evaporation, the crucible and contents should be heated from below. The use of top heating alone, as with a radiant heater, can result in incomplete elimination of silica by hydrofluoric acid.

Heat the crucible and residue, cautiously at first, over a gas flame and finally for 5 min at 1200°C. Cool and weigh. The difference between the two weights represents the "gravimetric" silica.

Preparation of the Solution for the Determination of Residual Silica, Titania, Alumina, Ferric Oxide, Manganese Oxide, and Phosphorus Pentoxide

NOTE: For aluminous materials this solution is also used for the determination of lime and magnesia.

Fuse the residue from the hydrofluoric acid treatment of the "gravimetric" silica with about 1 g of sodium carbonate and 0·2 g of boric acid and dissolve the melt in the main filtrate. If the residue is small it may be fused in 1–2 g of potassium pyrosulphate. Cool and dilute the solution to 500 ml in a volumetric flask. This solution is referred to as the "stock solution".

Determination of Residual Silica

The presence of phosphates, vanadates, etc., which produce a yellow colour with ammonium molybdate, will result in positive errors in the silica determination. Refractory materials rarely contain sufficient of these constituents to justify account being taken of them in the analytical procedure, unless a deliberate addition of phosphate has been made. The molybdenum blue method takes account of the presence of phosphate.

Molybdenum blue procedure

Transfer 5·0 ml of the stock solution to a 100-ml volumetric flask A, add 15 ml of water and 2 drops of 2,4-dinitrophenol indicator. Add diluted ammonia solution (1 + 1) dropwise until the indicator turns yellow (note the amount of ammonia used), then add 5 ml of diluted hydrochloric acid (1 + 4).

To another 100-ml volumetric flask B, add 20 ml of water and the same amount of diluted ammonia solution (1 + 1) as used to neutralize the aliquot in flask A. Add 2 drops of 2,4-dinitrophenol indicator followed by diluted hydrochloric acid (1 + 4) until the solution is neutral and then 5 ml in excess.

To both flasks add 6 ml of ammonium molybdate solution (80 g/litre) and stand for 5–10 min at a temperature of not less than 20°C and not greater than 30°C. Then add with swirling, 45 ml of diluted hydrochloric acid (1 + 1) and stand for 10 min.

Add 10 ml of stannous chloride solution (10 g/litre), dilute to 100 ml and mix. Measure the optical density of the solution in flask A against the solution in flask B in 10-mm cells at 800 nm, or by using a colour filter (Ilford 609) in a suitable instrument.

The colour is stable between 5 and 30 min after the addition of the stannous chloride solution.

Determine the silica content of the solution by reference to a calibration graph.

Alternative procedure (yellow silicomolybdic acid)

If phosphate is known not to be present in significant amounts, the following simpler alternative method may be used.

Transfer 20-ml aliquots of the stock solution to each of two 50-ml volumetric flasks, A and B, and add 2 drops of 2,4-dinitrophenol indicator. (If 10-ml aliquots are taken then the volume should be made up to 20 ml by the addition of 10 ml of water).

To flask B add ammonia solution ($d = 0.88$) dropwise until the indicator turns yellow (note the number of drops of ammonia solution added), then add 5 ml of diluted hydrochloric acid (1 + 4).

To flask A add 5 ml of diluted hydrochloric acid (1 + 4), followed by the same number of drops of ammonia solution ($d = 0.88$) as used to neutralize the solution in flask B. Then add 5 ml of ammonium molybdate solution (80 g/litre), slowly with swirling.

Dilute the solution in each flask to 50 ml with distilled water and shake well. Measure the silicomolybdate colour in flask A against the solution in flask B in 40-mm cells at 440 nm or by using a colour filter (Ilford 601) in a suitable instrument.

Measure the optical density between 5 and 15 min after the addition of the ammonium molybdate. Determine the silica content of the solution by reference to a calibration graph.

If the solution becomes cloudy on adding the ammonium molybdate or on standing, then the molybdenum blue method must be used.

Calculation of Total Silica Content

Add the residual silica content to the figure obtained for the "gravimetric" silica to obtain the total silica content.

D

Determination of Alumina

Separation of ferric oxide and titania

For high-silica materials, transfer a 200-ml aliquot of the stock solution to a 500-ml separating funnel, add 30 ml of hydrochloric acid ($d = 1\cdot18$), and 25ml of diluted sulphuric acid $(1 + 1)$.

For aluminosilicates and aluminous materials, transfer a 100-ml aliquot of the stock solution to a 500-ml separating funnel and add 10 ml of hydrochloric acid ($d = 1\cdot18$) and 25 ml of diluted sulphuric acid $(1 + 1)$. Where the lime content and lime-alumina ratio is low, the acidity may be adjusted by adding 20 ml of hydrochloric acid/100-ml aliquot and omitting the sulphuric acid addition.

To the appropriate solution add 20 ml of chloroform and 10 ml of cupferron solution (60 g/litre). Stopper the funnel and shake vigorously. Release the pressure in the funnel by carefully removing the stopper and rinse the stopper and neck of the funnel with water.

Allow the layers to separate and withdraw the chloroform layer. Confirm that extraction is complete by checking that the addition of a few drops of cupferron solution (60 g/litre) to the aqueous solution does not produce a permanent coloured precipitate.

Add further 10-ml portions of chloroform and repeat the extraction until the chloroform layer is colourless. Discard the chloroform extracts.

(*a*) *Volumetric determination of alumina* (*preferred method*). Run the aqueous solution from the cupferron–chloroform separation into a 500-ml conical flask. Add a few drops of bromophenol blue indicator and add ammonia solution ($d = 0\cdot88$) until just alkaline.

Re-acidify quickly with hydrochloric acid ($d = 1\cdot18$) and add 5–6 drops in excess.

Add sufficient standard EDTA solution (0·05M) to provide an excess of a few millilitres over the expected amount. Then add ammonium acetate buffer solution until the indicator turns blue followed by 10 ml in excess. Boil the solution for 10 min and cool rapidly.

Add an equal volume of ethanol (95%) and 1–2 ml of dithizone indicator followed by Naphthol Green B solution (1 g/litre) drop by drop until the reddish tint is screened out and titrate with standard zinc solution (0·05M) from blue/green to the first appearance of a permanent pink colour.

Calculation—If the EDTA solution is not exactly 0·05M, calculate the equivalent volume of 0·05M EDTA.

If V ml is the volume of EDTA (0·05M) and v ml is the volume of standard zinc solution (0·05M) used in the back-titration, then

$$Al_2O_3\ (\%) = 1\cdot275\ (V\text{-}v) \text{ for a 100-ml aliquot and}$$
$$Al_2O_3\ (\%) = 0\cdot6375\ (V\text{-}v) \text{ for a 200-ml aliquot.}$$

(*b*) *Gravimetric determination of alumina* (*alternative method*). Transfer the aqueous solution from the cupferron-chloroform separation to a 400- or 500-ml beaker. Boil for a few minutes to remove any traces of chloroform and then cool.

Neutralize with ammonia solution ($d = 0.88$) to the change-point of methyl red and then make just acid with hydrochloric acid ($d = 1.18$). With high-silica materials, in order to ensure full recovery of the small amount of aluminium normally present, add 10 ml of the standard aluminium solution (1.0 mg Al_2O_3/ml) before adjusting the acidity. For work of the highest accuracy, the aluminium content of 10 ml of the standard solution should be determined at the same time. Add 4 ml of diluted hydrochloric acid ($1 + 1$) in excess.

Add 2 ml of hydroxyammonium chloride solution (500 g/litre) and 2 ml of 1:10-phenanthroline solution (10 g/litre) and heat to 40–50°C. The volume of the solution at this stage should be about 350–400 ml (high-silica materials) or 250–300 ml (aluminosilicates and aluminous materials).

Add the requisite amount of 8-hydroxyquinoline solution (50 g/litre) and then add slowly, with stirring, 40 ml of ammonium acetate solution (approx. 40%). For solutions containing up to 30 mg of Al_2O_3, use 10 ml of 8-hydroxy-quinoline solution; for solutions containing 35–65 mg of Al_2O_3 use 20 ml of 8-hydroxyquinoline solution, for solutions containing 65–90 mg of Al_2O_3, use 30 ml of 8-hydroxyquinoline solution.

Heat, with stirring, to 60–65°C, and allow to stand at this temperature for 10 min stirring periodically. Allow to cool for 30 min.

Filter through a weighed crucible with a sintered-glass mat (porosity 4), scrubbing the beaker with a "bobby". Wash the precipitate with warm water (40–50°C).

Dry the crucible with the precipitate in an oven maintained at 150°C until constant weight is obtained; 2 h is normally sufficient.

For work of the highest accuracy, dissolve the precipitate by passing small quantities of nearly boiling, diluted hydrochloric acid ($1 + 1$) through the crucible and wash thoroughly with hot water. This method of acid extraction is to correct for small amounts of acid-insoluble material occasionally precipitated with the aluminium complex. Dry the crucible and residue at 150°C for 1 h, cool and weigh.

Calculation—Weight of aluminium oxinate $\times$ 0.1110 = weight of Al_2O_3.

Determination of Ferric Oxide

Transfer a 5-ml aliquot of the stock solution to a 100-ml volumetric flask. Add 2 ml of hydroxyammonium chloride solution (100 g/litre), 5 ml of 1:10-phenanthroline solution (10 g/litre) and 2 ml of ammonium acetate solution (approx. 10%).

Allow to stand for 15 min, dilute to 100 ml and mix.

Measure the optical density of the solution against water in 10-mm cells at 510 nm or by using a colour filter (Ilford 603) in a suitable instrument. The colour is stable between 15 and 75 min after the addition of the ammonium acetate solution.

Determine the ferric oxide content of the solution by reference to a calibration graph.

Using the Unicam SP600 spectrophotometer, the calibration graph extends to about 5% Fe_2O_3. If this amount is exceeded, it is permissible to dilute the stated aliquot to 200 ml in a volumetric flask instead of to 100 ml.

Determination of Titania

Transfer a 20-ml aliquot of the stock solution to each of two 50-ml volumetric flasks, A and B. To each flask add 10 ml of diluted phosphoric acid (2 + 3) and, to flask A only, 10 ml of hydrogen peroxide solution (6%). Dilute the solution in each flask to 50 ml and shake well.

Measure A against B in 40-mm cells at 398 nm or by using a colour filter (Ilford 601) in a suitable instrument. The colour is stable between 5 min and 24 h after the addition of the hydrogen peroxide solution.

Determine the titania content of the solution by reference to a calibration graph.

Determination of Manganese Oxide

Transfer a 50-ml aliquot of the stock solution to a 250-ml beaker. Add 10 ml of diluted sulphuric acid (1 + 1), 5 ml of nitric acid ($d = 1.42$) and evaporate to strong fumes to remove chlorides.

Cool, add 20 ml of nitric acid ($d = 1.42$), 10 ml of diluted phosphoric acid (1 + 9) and about 50 ml of water.

Boil to dissolve salts and to remove nitrous fumes, cool slightly and add about 0.2 g of potassium periodate.

Boil for 2 min, transfer to a steam bath and keep hot for 10 min.

Allow to cool and transfer to a 100-ml volumetric flask. Dilute the solution in the flask to 100 ml and mix. Measure the optical density of the solution against water in 40-mm cells at 524 nm, or by using a colour filter (Ilford 604) in a suitable instrument.

Determine the manganese oxide content of the solution by reference to a calibration graph.

Determination of Phosphorus Pentoxide

Transfer 50-ml aliquots of the stock solution to each of two 250-ml beakers. To each add 10 ml of nitric acid ($d = 1.42$), 1 ml of diluted sulphuric acid (1 + 1) and evaporate to dryness, being careful to avoid spurting.

Cool, dissolve the residues in 10 ml of diluted nitric acid (1 + 2) and transfer to two 100-ml volumetric flasks, A and B, filtering if necessary.

To each flask add 10 ml of ammonium vanadate solution (2.5 g/litre) and to flask A only add 6 ml of ammonium molybdate solution (80 g/litre). Dilute the solution in each flask to 100 ml and shake well.

Measure the phosphovanadomolybdate colour in flask A against the solution in flask B in 40-mm cells at 450 nm, or by using a colour filter (Ilford 601) in a suitable instrument. The colour is stable for up to 30 min after the addition of the ammonium molybdate solution.

Determine the phosphorus pentoxide content of the solution by reference to a calibration graph.

Determination of Lime, Magnesia and Alkalis

Decomposition of the Sample

Weigh the finely ground, dried (110°C) sample into a small platinum basin (0·250 g for aluminous materials and aluminosilicates; 1·00 g for high-silica materials) and ignite gently to remove organic matter.

To the cool dish add 10 ml of sulphuric–nitric acid mixture and about 10 ml of hydrofluoric acid (40 % w/w). Transfer the vessel to a sand bath in a fume cupboard, allow to react thoroughly with the lid on for about 15 min, then evaporate to dryness, being careful to avoid spurting.

Cool, add 10 ml of the sulphuric-nitric acid mixture, and rinse down the sides of the basin with water. Evaporate carefully to dryness.

To the cool, dry residue add 20 ml of diluted nitric acid (1 + 19) and warm to dissolve.

Cool, filter if necessary, and dilute the solution to 250 ml in a volumetric flask, to form the sample solution A.

NOTE: As sulphuric acid is used for the decomposition it may be preferable to determine lime and magnesia on the main stock solution if these constituents are present in large quantities.

Determination of Lime

For high-silica materials and aluminosilicates, pipette an aliquot of the sample solution A prepared by the hydrofluoric acid treatment (100 ml for aluminosilicates, 50 ml for high-silica materials) into a 500-ml conical flask.

Add 5 ml of diluted triethanolamine (1 + 1) and 10 ml of potassium hydroxide solution (250 g/litre) and dilute to about 200 ml.

For aluminous materials, pipette a 50-ml aliquot of the stock solution (prepared from the silica filtrate and the silica residue) into a 500-ml conical flask. Add 5 ml of diluted triethanolamine (1 + 1) and 20 ml of potassium hydroxide solution (250 g/litre) and dilute to about 200 ml.

To the appropriate solution add about 0·015 g of screened Calcein indicator and titrate with standard EDTA solution (5 g/litre) from a semi-micro burette, the colour change being from fluorescent green to pink. The end-point is the final change in colour.

Determination of the Sum of Lime and Magnesia

For high-silica materials and aluminosilicates, pipette an aliquot of the sample solution A prepared by the hydrofluoric acid treatment (100 ml for aluminosilicates, 50 ml for high-silica materials) into a 500-ml conical flask.

Add 10 drops of hydrochloric acid ($d = 1·18$), 20 ml of diluted triethanolamine (1 + 1) and 25 ml of ammonia solution ($d = 0·88$) and dilute to about 200 ml.

For aluminous materials, pipette a 50-ml aliquot of the stock solution (prepared from the silica filtrate and the silica residue) into a 500-ml conical flask. Add 20 ml of diluted triethanolamine (1 + 1) and 25 ml of ammonia solution ($d = 0·88$) and dilute to about 200 ml.

To the appropriate solution add about 0·04 g of methylthymol blue complexone indicator and titrate with standard EDTA solution (5 g/litre) from a semi-micro burette, the colour change being from blue to colourless.

Calculation of Magnesia

The volume of EDTA used for the titration of lime is subtracted from the volume of EDTA used for the titration of the sum of lime and magnesia. The remainder represents the volume of EDTA required for the titration of the magnesia.

Determination of Alkalis

Determine the alkalis in the sample flame-photometrically by the method described in Chapter 16 under "High-silica, Aluminosilicate and Aluminous Materials".

B. Single-Dehydration Method

Where it is not desired to use the coagulation method for the determination, of silica, the following alternative method is allowed in B.S. 1902. Loss on ignition is carried out as described previously.

DECOMPOSITION OF THE SAMPLE

NOTE: For routine work, the loss on ignition may be carried out in the platinum dish and the same portion of the sample used for the fusion. The technique is only admissible if the amount of sintering of the sample after the determination of loss on ignition does not prevent adequate mixing of the sample and flux.

High-silica Materials and Aluminosilicates

Procedure

Weigh 1·000 g of the finely ground, dried (110°C) sample into a platinum dish, a convenient size is 70-mm dia., 40-mm deep, of effective capacity 75 ml. Add 3 g of fusion mixture and mix intimately to form a charge of about 50-mm dia. in the centre of the dish.

Heat over a mushroom burner, cautiously at first, then gradually raise the temperature to the full heat of the burner over a period of about 10 min, the dish being covered with a lid to prevent loss by spurting and to reduce heat losses. In some cases it may be necessary to transfer the dish to a Méker burner to melt the flux effectively.

Finally, heat the dish in a muffle furnace at 1200°C for 5 min. Remove the dish from the furnace and allow to cool.

If the fusion appears blue or blue-green indicating the presence of maganese, add 2 ml of ethanol (95%) to the contents of the dish prior to dissolution.

Add 30 ml of diluted hydrochloric acid (1 + 1) and 1 ml of diluted sulphuric acid (1 + 1) and allow the covered dish to stand on a steam bath to facilitate disintegration of the melt.

When disintegration is apparently complete, crush with the aid of a glass rod any large flakes of insoluble material.

Aluminous Materials

Additional Reagent

Sorbitol

Prepared Reagent

Sorbitol: Dissolve 20 g of sorbitol in water, dilute to 300 ml and mix. This solution will keep for about 2 weeks.

Procedure

Weigh 1·000 g of the finely ground, dried (110°C) sample into a platinum dish, a convenient size is 70-mm dia., 40-mm deep, of effective capacity 75 ml. Add 1·5 g of fusion mixture and 0·2 g of boric acid and mix intimately to form a charge of about 50-mm dia. in the centre of the dish. Cover the dish with a lid.

Heat over a gas burner for about 5 min; it is unnecessary to take further precautions to prevent spurting. Finally, heat the dish in a muffle furnace at 1200°C for 30 min. Remove the dish from the furnace and cool.

Add 15 ml of the sorbitol solution, 5 ml of sulphuric acid (1 + 9) and 15 ml of hydrochloric acid ($d = 1·18$); if the presence of manganese is suspected from the colour of the fusion, add 2 ml of ethanol (95%). Cover the dish with a lid and simmer gently until the melt is completely decomposed. If necessary, aid the decomposition by carefully stirring with a glass rod. When disintegration of the melt is complete, scrub the lid with a rubber-tipped glass rod and allow the washings to run into the dish.

DETERMINATION OF THE MAIN SILICA

All Materials

Transfer the dish and contents to a steam bath and evaporate the solution to apparent dryness. Allow the dish to remain on the steam bath until the smell of hydrochloric acid can no longer be detected. The evaporation can be hastened by the occasional use of a glass rod to break up the gel that tends to slow down the evaporation. For aluminous materials transfer the dish to an oven and bake at 110°C for 30 min (the presence of sorbitol may result in a small amount of charring which discolours the silica precipitate; this organic matter burns off during ignition and is of no significance). Remove from the oven and cool.

Moisten the contents by adding 10 ml of hydrochloric acid ($d = 1·18$). After about 1 min, add 25 ml of hot water and stir the mixture for a few moments until solution of soluble salts appears to be complete; then digest the mixture on a steam bath for 10 min. A Whatman accelerator tablet may be added at this stage to increase the speed of filtration. Filter through a 90-mm No. 42 Whatman paper and transfer the silica to the filter with hot hydrochloric acid (1 + 19), scrubbing the dish with a rubber-tipped glass rod. Wash the precipitate 5 times with hot hydrochloric acid (1 + 19) and then 24 times with hot water. Reserve the filtrate and washings.

Transfer the paper and precipitate to an ignited and weighed platinum crucible. Ignite at a low temperature until the precipitate is free from carbonaceous matter and then heat in a muffle furnace at 1200°C to constant weight, 30 min being normally sufficient.

Moisten the contents of the cold crucible with water, add 5 drops of diluted sulphuric acid (1 + 1) and 10 ml of hydrofluoric acid (40% w/w). Evaporate to dryness on a sand bath in a fume cupboard (for the evaporation, the crucible and contents should be heated from below: the use of top heating alone, as with a radiant heater, can result in incomplete elimination of silica by the hydrofluoric acid.)

Heat the crucible and residue, cautiously at first, over a gas flame and finally for 5 min at 1200°C, cool and weigh. If the residue weighs more than 5 mg, repeat the treatment with sulphuric and hydrofluoric acids to ensure that all the silica is removed. The difference between the two weights represents the "gravimetric" silica.

PREPARATION OF THE SOLUTION FOR THE DETERMINATION OF RESIDUAL SILICA, FERRIC OXIDE, TITANIA, MANGANESE OXIDE, PHOSPHORUS PENTOXIDE AND ALUMINA

NOTE: For aluminous materials this solution is also used for the determination of lime and magnesia.

Fuse the residue from the hydrofluoric acid treatment of the "gravimetric" silica with 1 g of potassium pyrosulphate, dissolve the melt in water containing a few drops of hydrochloric acid ($d = 1.18$) and add the solution to the main filtrate. Cool, dilute the combined solution to 500 ml in a volumetric flask and mix.

This solution is referred to as the "stock solution".

DETERMINATION OF RESIDUAL SILICA

The presence of phosphates, vanadates, etc., which produce a yellow colour with ammonium molybdate, will result in positive errors in the silica determination. Normal refractory materials rarely contain sufficient of these constituents to justify account being taken of them in the analysis procedure. The molybdenum blue method takes account of the presence of phosphate.

Procedure

Phosphate present: Transfer a 10·0-ml aliquot of the "stock solution" to a 100-ml volumetric flask A, add 10 ml of water and 2 drops of 2:4 dinitrophenol indicator. Add diluted ammonia solution (1 + 1) dropwise until the indicator turns yellow (count the number of drops of ammonia used), then add 5 ml of diluted hydrochloric acid (1 + 4).

To another 100-ml volumetric flask B, add 20 ml of water and the same amount of diluted ammonia solution (1 + 1) as used to neutralize the aliquot in flask A. Add 2 drops of 2:4 dinitrophenol indicator followed by diluted hydrochloric acid (1 + 4) until the indicator turns colourless, then add 5 ml of diluted hydrochloric acid (1 + 4) in excess.

To both flasks, A and B, add 6 ml of ammonium molybdate (80 g/litre) and stand for 5–10 min at a temperature of 20–30°C. Then add, with swirling, 45 ml of diluted hydrochloric acid (1 + 1) and stand for 10 min. Add 10 ml of stannous chloride solution (10 g/litre), dilute to 100 ml and mix.

Measure the optical density of the solution in flask A against the solution in flask B in 10-mm cells at 800 nm or by using a colour filter (Ilford 609) in a suitable instrument. The colour is stable for 5–30 min after the addition of the stannous chloride solution.

Determine the silica content of the solution by reference to a calibration graph.

Phosphate absent: transfer 20·0 ml, or 10·0 ml (+ 10 ml of water) if the sample has a high titania content, of the "stock solution" to each of two 50-ml volumetric flasks, A and B, and neutralize the excess of acid in each flask by the dropwise addition of diluted ammonia solution (1 + 1) until the first appearance of a permanent precipitate. Add immediately 5 ml of diluted hydrochloric acid (1 + 4) and, to flask A only, 5 ml of ammonium molybdate solution (80 g/litre).

Dilute the solution in each flask to 50 ml and shake well. Measure the silicomolybdate colour in flask A against the solution in flask B in 40-mm cells at 440 nm or by using a colour filter (Ilford 601) in a suitable instrument.

Measure the optical density not sooner than 5 min and not later than 15 min after the addition of the ammonium molybdate solution.

Determine the silica content of the solution by reference to a calibration graph.

Calculation of Total Silica Content

Add the residual silica content to the figure obtained for the "gravimetric" silica to obtain the total silica content.

DETERMINATION OF ALUMINA, FERRIC OXIDE, TITANIA, MANGANESE OXIDE AND PHOSPHORUS PENTOXIDE

Aliquots of the stock solution are used for the above determinations by the methods described on pages 50–52.

NOTE: For aluminous materials this solution is also used for the determination of lime and magnesia, by the methods described on pages 53-54.

D*

CHAPTER 6

ANALYSIS OF HIGH-SILICA MATERIALS, ALUMINOSILICATES AND ALUMINOUS MATERIALS:

CLASSICAL METHOD

GENERAL

This method should be reserved for those materials to which the coagulation or the single-dehydration method cannot be applied. This can arise because of serious deviation from the normal composition, either by the presence of an abnormal amount of one constituent or by the presence of constituents which are not allowed for in the other accurate methods. In the case of the single-dehydration method the presence of a large amount of titania can result in hydrolysis which gives a cloudy filtrate from the silica determination, precluding the use of spectrophotometric methods except by precipitation of the ammonia group oxides, but this is overcome by the use of the coagulation technique. Also manganese oxide, if present in amounts greater than about 0·25 %, causes a green coloration in the lime and magnesia solutions which destroys the end-points. This can of course be overcome by carrying out a DDC extraction as described under the determination of magnesia in magnesites (p. 110), but may in general be overcome by using the classical method.

If elements other than those catered for in the coagulation and single-dehydration methods are present, the more flexible classical method is often preferable, modifications being made as required by the need to carry out additional determinations. Some of these modifications are dealt with in Chapter 15 on the analysis of frits and glazes. Other modifications must be based on a general knowledge of the analytical chemistry of the additional constituents.

It should be borne in mind, however, that more modern techniques are available to overcome many problems which, in the past, have necessitated recourse to the classical method. A number of suggestions of this type of new approach are provided in Chapter 15. This sort of approach has been made to many problems in the Association's laboratory but investigations have rarely been carried out in adequate detail to justify a detailed method.

As indicated above, the procedure in this chapter only includes those elements normally analysed for in these classes of material.

PRINCIPLE OF THE METHOD

The loss on ignition is determined by heating the sample to 1000°C for 30 min. The main portion of the analysis is carried out by fusing either the ignited portion or a freshly weighed portion of the sample in either fusion mixture (samples containing less than 45 % Al_2O_3) or in fusion mixture/boric acid flux (aluminous materials). The modern deviant of the method includes a double dehydration of silica in the same 70-mm platinum dish as is used for the fusion. This makes for more effective dehydration, is without any risk of loss due to transference, and allows the precipitate to be more readily "bobbied out". The use of a relatively low flux-to-sample ratio minimizes subsequent washing of precipitates and allows the use of the smaller dishes (cf. 125-mm porcelain basins).

The melt is dissolved in hydrochloric acid and the solution evaporated to dryness on a steam bath. The residue is dissolved in hydrochloric acid and water, and the dehydrated silica filtered off. The filtrate is again dehydrated, the residue redissolved and a second silica precipitate filtered off. The combined precipitates are ignited and weighed, before and after hydrofluoric acid treatment. The residue from this treatment is fused in sodium carbonate, the melt dissolved in diluted hydrochloric acid and the solution added to the filtrate from the "second" silica.

The ammonia group oxides are determined by precipitation with ammonia in the presence of ammonium chloride and the resulting hydroxides of iron, titanium and aluminium filtered off. These are redissolved in acid and reprecipitated, to minimize co-precipitation. The acidified combined filtrates are reduced in volume and a "catch" precipitation with ammonia carried out. It is usual to oxidize any manganese with bromine to ensure its precipitation as a hydrated MnO_2. The purpose of this precipitation is to ensure that any alumina still in the solution is finally precipitated. This alumina can have escaped precipitation by a failure to achieve the correct pH on either of the two main precipitations (the presence of a large amount of iron may colour the precipitate so much that the colour of the indicator is difficult to see). Alternatively, the large amount of wash liquor may have caused some redissolution, or some precipitate may have "crept" or even accidently washed over the paper.

The combined precipitates are ignited (for accurate work a temperature of 1200°C is essential), weighed, fused in sodium carbonate/boric acid, dissolved in sulphuric acid and made up to volume. Iron, titanium and manganese are determined spectrophotometrically on this solution with 1:10 phenanthroline, hydrogen peroxide, and potassium periodate respectively. The original version of this method then allowed the calculation of alumina by difference, but in the current version its actual determination by either oxine or EDTA is proposed. The analyst must note in applying this method that some silica will be entrained in the R_2O_3 precipitate and by applying the alkali-fusion technique this may be determined colorimetrically. The alternative acid-fusion method involves a dehydration of this solution to remove the bulk of this silica. It should also be borne in mind that other elements are included in the R_2O_3 precipitate, the most obvious of these being phosphorus and zirconium, but a number of other less common elements may also be precipitated.

The filtrate from the ammonia precipitation is acidified, reduced in volume and the lime precipitated as the oxalate. Reprecipitation is again necessary to ensure the purity of the precipitate. The second precipitate is redissolved with sulphuric acid and the oxalate titrated with permanganate. If larger amounts of lime are present it is often preferable to ignite the precipitate to 1000°C and weigh the CaO. If the lime/magnesia ratio is unfavourably high a third precipitation may prove necessary.

The combined filtrates from the lime precipitations are acidified, reduced in volume, and the magnesia precipitated as phosphate. This precipitate is redissolved and reprecipitated to ensure not only purity but the correct form, so that ignition is to the pyrophosphate. The precipitate is stood for some time at a low temperature (the solubility increasing with temperature), filtered off and ignited. The ignition temperature of 1150°C is selected to be the lowest temperature at which the magnesium pyrophosphate will burn white, at any lower temperature it will be grey in colour, and the higher the temperature the greater the risk of damage to the platinum by the phosphate.

In the procedure given here, alkalis are determined flame-photometrically in a separate portion of sample decomposed with hydrofluoric acid. Strictly, the "classical" method should include the use of the "Lawrence Smith" method, but this is now so outdated that its inclusion is not justified.

REAGENTS

Additional Reagents

Barium diphenylamine sulphonate
Mercuric chloride
Sorbitol (for aluminous materials)
Tin: granulated

Prepared Reagents

Ammonium acetate (approx. 40%): Dilute 570 ml of acetic acid (glacial) to 1700 ml with water and add carefully 570 ml of ammonia solution ($d = 0.88$). Mix, cool and adjust to pH 6·4–6·8, either with acetic acid or ammonia solution.

Ammonium acetate (approx. 10%): Dilute the ammonium acetate solution (approx. 40%) with three times its volume of water and mix.

Ammonium acetate buffer: Add 120 ml of acetic acid (glacial) to 500 ml of water followed by 74 ml of ammonia solution ($d = 0.88$). Mix, cool and dilute to 1 litre.

Ammonium nitrate (approx. 1%): Dilute 10 ml of nitric acid ($d = 1.42$) to about 200 ml. Add diluted ammonia solution (1 + 1) until the solution is faintly alkaline to methyl red. Dilute to 1 litre.

Ammonium phosphate (100 g/litre): Dissolve 2 g of di-ammonium hydrogen orthophosphate in 20 ml of water. This solution should be freshly prepared

Bromine water (saturated): Shake 20 ml of bromine with 500 ml of water in a glass-stoppered bottle.

Cupferron (60 g/litre): Dissolve 3 g cupferron in 50 ml of cold water and filter. This solution should be freshly prepared. If the reagent is discoloured or gives a strongly coloured solution a new stock should be obtained. The solid reagent should be stored in a tightly stoppered bottle in the presence of a piece of ammonium carbonate to prevent decomposition.

Hydrogen sulphide wash solution: Add 20 ml of hydrochloric acid ($d = 1·18$) to 980 ml of water and saturate with hydrogen sulphide.

8-Hydroxyquinoline (50 g/litre): Dissolve, by warming, 50 g of 8-hydroxyquinoline in 120 ml of acetic acid (glacial). Dilute to about 700 ml, filter, cool and dilute to 1 litre.

Mercuric chloride (saturated): Shake 20 g of mercuric chloride with 250 ml of water. Allow to stand 24 h before use.

1:10 Phenanthroline (10 g/litre): Prepare enough solution for immediate use at a rate of 0·1 g of 1:10 phenanthroline hydrate in 10 ml of diluted acetic acid (1 + 1).

Silver nitrate (1 g/litre)—used for testing for chlorides: Dissolve 0·1 g of silver nitrate in 100 ml of diluted nitric acid (1 + 9).

Sorbitol: Dissolve 20 g of sorbitol in water, dilute to 300 ml and mix. This solution will keep for about 2 weeks.

Stannous chloride (30 g/litre): Dissolve, by warming, 3 g of stannous chloride in 30 ml of hydrochloric acid ($d = 1·18$). Cool and dilute to 100 ml. Store in a rubber-stoppered bottle, with a small piece of tin in the solution to allow for oxidation.

Sulphuric-nitric acid mixture: To 650 ml of water, add 100 ml diluted sulphuric acid (1 + 1) and 250 ml of nitric acid ($d = 1·42$).

Sulphuric-phosphoric acid mixture: Cautiously add 75 ml of sulphuric acid ($d = 1·84$) and 75 ml of phosphoric acid ($d = 1·75$) to 350 ml of water, cooling the solution and keeping it well mixed during the addition of the acids.

Indicators

Barium diphenylamine sulphonate
Bromophenol blue
Dithizone
Methyl red

Standard Solutions

Aluminium (1·0 mg Al_2O_3/ml)
EDTA (0·05M)
Potassium dichromate (0·05N)
Potassium permanganate (approx. 0·1N)
Zinc (0·05M)

PREPARATION OF SAMPLE

The sample prepared for analysis should be ground to pass completely through a 120-mesh B.S. test sieve. A non-metallic (e.g. nylon bolting-cloth sieve is preferable).

High-silica materials and aluminosilicates containing up to 45 % of alumina may be ground in agate, or in iron with subsequent treatment with a magnet.

Aluminous materials may be ground in alumina or in iron mortars: in the latter case the ground sample should be treated with a magnet. It may be necessary to prepare a sample by each method and carry out the main analysis on the iron-ground sample with a ferric oxide determination on the sample ground in alumina, and then calculate the results to the latter iron content. Any metallic iron introduced will be oxidized to the ferric state by the heat treatment during the determination of loss of ignition. Correction must be made to this figure for the oxygen taken up and the results should be corrected for the difference in ferric oxide content of the two samples, expressed as metallic iron, the state in which it is present when weighing out the sample. The use of a boron carbide mortar permits the grinding of aluminous materials with the minimum of contamination, thus obviating the necessity for preparing two separate portions of the sample.

PROCEDURE

Determination of Loss on Ignition

Weigh 1·000 g of the finely ground, dried (110°C) sample into a platinum crucible. Place the crucible in a muffle furnace and slowly raise the temperature to 1000 ± 25°C. Ignite to constant weight at this temperature; 30-min ignition is usually sufficient.

For routine work it is possible to start the ignition over a low mushroom flame, slowly increasing the temperature to full heat over a period of about 20 min, after which the crucible is transferred to a furnace at 1000°C for 30 min.

Determination of Silica, Alumina, Ferric Oxide, Titania, Manganese Oxide, Lime and Magnesia

Decomposition of the Sample

High-silica materials and aluminosilicates

Weigh 1·000 g of the finely ground, dried (110°C) sample into a platinum crucible or dish; a suitable size for the dish is 70-mm dia. and 40-mm deep, of effective capacity 75 ml.

Add 3 g of fusion mixture and mix intimately.

For routine work, the loss on ignition may be carried out in the dish and the same portion of sample used for the fusion. This technique is only admissible if the amount of sintering of the sample after the determination of loss on ignition does not prevent adequate mixing of the sample and flux.

Heat over a mushroom burner, cautiously at first, then gradually raise the temperature to the full heat of the burner over a period of about 10 min, the dish or crucible being covered with a lid to prevent loss by spurting and to reduce heat losses. In some cases it may be necessary to transfer the dish to a Méker burner to melt the flux effectively.

Finally heat the dish or crucible in a muffle furnace at 1200°C for 5 min. Remove the vessel from the furnace and allow to cool.

If the fusion has been carried out in a dish, add 30 ml of diluted hydrochloric acid (1 + 1) and allow the covered dish to stand on a steam bath to facilitate disintegration of the melt. When disintegration is apparently complete, crush with the aid of a glass rod any large flakes of insoluble material.

If the fusion has been carried out in a crucible and it is possible to detach the main portion of the melt, transfer this to a platinum or porcelain basin, and cover with a clock glass. Dissolve the residue of the melt remaining in the crucible by warming with a little diluted hydrochloric acid (1 + 1) and "bobby" the crucible, adding the solution and washings to the main portion in the basin. Add 20 ml of hydrochloric acid ($d = 1 \cdot 18$) and about 40 ml of hot water and warm on a steam bath until the melt is completely disintegrated. Crush any lumps remaining in the solution.

If the melt cannot be detached from the crucible, place the crucible and lid in a 250-ml beaker and add about 100 ml of hot water. Cover the beaker with a clock glass and carefully add 20 ml of hydrochloric acid ($d = 1 \cdot 18$). Warm until disintegration of the melt is complete and remove the crucible and lid, washing them thoroughly and scrubbing them with a "bobby". Crush any lumps remaining in the solution and wash out the beaker with hot water into a porcelain evaporating basin scrubbing out the beaker with a "bobby".

Aluminous materials

Weigh 1·000 g of the finely ground, dried (110°C) sample into a platinum crucible or dish, a convenient size for the dish is 70-mm dia., 40-mm deep, of effective capacity 75 ml. Add 1·5 g of fusion mixture and 0·2 g of boric acid and mix intimately. Decompose the sample as above except that the fusion is continued for 30 min at 1200°C.

Add 15 ml of the sorbitol solution, 5 ml of diluted sulphuric acid (1 + 9) and 15 ml of hydrochloric acid ($d = 1 \cdot 18$). If the presence of manganese is suspected from the blue or green-blue colour of the fusion, add 2 ml of ethanol (95%). Cover the dish with a lid and simmer gently until the melt is completely decomposed. Scrub the lid with a rubber-tipped glass rod and allow the washings to run into the dish.

If the fusion has been carried out in a crucible, treat as above (aluminosilicates) but using the reagents for aluminous materials.

Determination of Silica

All materials

Evaporate the solution from the fusion to dryness on a steam bath, from time to time breaking up the crust that forms and hinders evaporation.

When the residue is completely dry, cover the basin with a clock glass and drench the residue with 10 ml of hydrochloric acid ($d = 1 \cdot 18$). Allow to stand for a few minutes, add about 35 ml of hot water and digest on a steam bath for 5 min to dissolve the salts, then filter through a No. 41 Whatman paper.

Transfer the silica to the filter with a jet of hot, diluted hydrochloric acid (1 + 19) and scrub the basin with a "bobby". Wash the residue five times with hot, diluted hydrochloric acid (1 + 19) and then with hot water until it is free from chlorides.

Reserve the residue and paper for the subsequent ignition and transfer the filtrate and washings back to the evaporating basin. Again evaporate completely to dryness on a steam bath, cover the basin with a clock glass and bake in an air oven at 110°C for 1 h. Allow to cool, then drench the residue with 10 ml of hydrochloric acid ($d = 1.18$). Allow to stand for a few minutes, add about 75 ml of hot water and digest for 5 min on a steam bath to dissolve the salts. Filter through a No. 42 Whatman paper, transferring the residue to the paper with a jet of hot, diluted hydrochloric acid (1 + 19) and scrubbing the basin with a "bobby".

Wash five times with hot, diluted hydrochloric acid (1 + 19) and then with hot water until the residue is free from chlorides.

Reserve the filtrate and washings for the determination of the ammonia group oxides.

Place the two residues and papers, without drying, in a weighed platinum crucible and heat cautiously to dry the residue and char the papers. Then burn off the carbon at a low temperature, otherwise it may be impossible to remove all of it, and finally ignite at 1200°C for 30 min and then to constant weight. Allow to cool, and weigh to obtain the weight of the impure silica.

Moisten the residue with 10 drops of diluted sulphuric acid (1 + 1) and about 10 ml of hydrofluoric acid (40% w/w). Evaporate slowly to dryness on a sand bath in a fume cupboard.

Ignite the dry residue at 1200°C for 5 min, allow the crucible to cool and weigh.

Subtract the weight of this residue from the weight of the impure silica to obtain the weight of the silica in the sample taken. A little silica may remain in the filtrate and this is recovered from the ammonia group oxides.

If the weight of the residue is more than 10 mg, repeat the treatment with sulphuric and hydrofluoric acids to ensure that all the silica has been removed. If the residue after this treatment is more than 10 mg, the determination should be regarded with suspicion, and if accurate results are required it should be repeated.

Fuse the residue with about 1 g of anhydrous sodium carbonate and dissolve the melt in diluted hydrochloric acid (1 + 1). Add the solution to that reserved for the determination of the ammonia group oxides.

Determination of the Total Ammonia Group Oxides (R_2O_3)

For work of the highest accuracy, dissolved platinum should be removed from the combined solutions containing the ammonia group oxides. Precipitate the platinum as sulphide by passing hydrogen sulphide through the solution, stand for about 1 h and filter. Wash with hydrogen sulphide wash solution and boil the filtrate to expel hydrogen sulphide.

Add 5 ml of bromine water (saturated) and boil for a few minutes to oxidize any ferrous salts. To the solution (approx. 300 ml) add 2–3 g of ammonium chloride, warm the solution to about 80°C and add diluted ammonia solution (1 + 1), with stirring, until the solution is just alkaline to bromophenol blue.

Boil the alkaline solution for 2 min and allow to stand for 5 min for the precipitate to settle.

High-silica materials

Filter through a No. 40 Whatman paper, transfer the precipitate to the filter and wash five times with hot, faintly ammoniacal ammonium nitrate solution (approx. 1 %). Just acidify the filtrate and washings with hydrochloric acid ($d = 1·18$) and reserve. Dissolve the precipitate through the filter by the dropwise addition of hot, diluted hydrochloric acid (1 + 1) into the precipitation beaker and wash the paper thoroughly with hot water.

Boil the solution to ensure complete dissolution and cool to about 80°C. To the solution (approx. 300 ml) add 2–3 g of ammonium chloride and then add diluted ammonia solution (1 + 1), with stirring, until the solution is just alkaline to bromophenol blue.

Boil the alkaline solution for 2 min, allow to stand for 5 min for the precipitate to settle and filter through the No. 40 Whatman paper used for the first filtration. Transfer the precipitate to the filter with a jet of hot, faintly ammonical ammonium nitrate solution (approx. 1 %), scrubbing the beaker with a "bobby".

Wash the precipitate free from chlorides with hot, faintly ammoniacal ammonium nitrate solution (approx. 1 %). Just acidify the filtrate and washings with hydrochloric acid ($d = 1·18$) and add to the solution reserved for the determination of lime and magnesia.

Aluminosilicates and aluminous materials

Filter through a No. 41 Whatman paper, transfer the precipitate to the filter and wash five times with hot, faintly ammoniacal ammonium nitrate solution (approx. 1 %). Just acidify the filtrate and washings with hydrochloric acid ($d = 1·18$) and reserve; transfer the precipitate and paper back to the precipitation beaker.

Dissolve the precipitate in a slight excess of hydrochloric acid ($d = 1·18$) and macerate the filter paper. Dilute to about 300 ml with water, warm the solution to about 80°C and add diluted ammonia solution (1 + 1), with stirring, until the solution is just alkaline to bromophenol blue.

Boil the alkaline solution for 2 min, allow to stand for 5 min to allow the precipitate to settle and filter through a No. 541 Whatman paper. Transfer the precipitate to the filter, scrubbing the beaker with a "bobby". Wash the precipitate free from chlorides with hot, faintly ammoniacal ammonium nitrate solution (approx. 1 %). Just acidify the filtrate with hydrochloric acid ($d = 1·18$), add to the reserved solution and retain the precipitate and paper for ignition.

Evaporate the combined filtrates to about 250 ml and add 5 ml of bromine water (saturated). Make just alkaline with diluted ammonia solution (1 + 1) and boil off the excess ammonia. Any slight precipitate should be filtered off on a No. 40 Whatman paper, washed as before and reserved for ignition. If any appreciable amount of precipitate is formed, reprecipitate (adding bromine water), filter and wash as before. Acidify the filtrate and washings with hydrochloric acid ($d = 1\cdot18$) and reserve for the determination of lime.

All materials

Place the precipitates and papers, reserved for ignition, in a weighed platinum crucible and heat, slowly at first, to dry the precipitates and char the papers. Burn off the carbon at a low temperature and finally ignite at 1200°C to constant weight.

The weight of the ammonia group oxides is thus obtained.

Dissolution of Ammonia Group Precipitate

Alkali fusion method (preferred)

Mix the precipitate with 5 g of sodium carbonate and 2 g of boric acid and fuse the residue, if necessary using a 1200°C furnace. Dissolve the cooled melt in about 150 ml of water to which 13 ml of sulphuric acid ($d = 1\cdot84$) has been cautiously added. Dilute the solution to 500 ml in a volumetric flask, to form the "stock test solution".

Acid fusion method (alternative)

Carefully fuse the ignited, weighed oxides in the same platinum crucible with about 8 g of potassium pyrosulphate. Cool the crucible and dissolve the melt in a beaker containing about 150 ml of water to which 10 ml of sulphuric acid ($d = 1\cdot84$) has been cautiously added.

Determination of Residual Silica

If the alkali fusion method has been used the residual silica may be determined on an aliquot of the "stock test solution" by either the yellow silicomolybdate or the molybdenum blue methods described on pages 56-57.

If the acid fusion method has been used the residual silica may be determined as follows:

Evaporate the solution from the pyrosulphate fusion to strong fumes. Cool, dilute with water to about 150 ml, warm to dissolve the salts and filter through a No. 42 Whatman paper. Wash eight times with hot water; reserve the filtrate. Ignite the residue at 1000°C for 10 min, cool and weigh.

Add 5 drops of diluted sulphuric acid (1 + 1) and 2–3 ml of hydrofluoric acid (40% w/w) and evaporate to dryness. Ignite the residue carefully at first, to prevent spurting, and finally at 1000°C for 5 min. Cool and weigh. The difference in weight gives the amount of residual silica.

Add the weight of this silica to the weight of silica obtained by dehydration and make a corresponding deduction from the weight of the ammonia group oxides.

Fuse the residue remaining in the crucible after the hydrofluoric acid treatment with 2 g of potassium pyrosulphate and dissolve the melt in the filtrate from the separation of residual silica.

Cool and dilute the solution of the pyrosulphate melts to 500 ml in a volumetric flask. The solution thus obtained will be referred to as the "stock test solution".

Determination of Ferric Oxide

If a spectrophotometer (e.g. Unicam SP600) is available, the colorimetric method is suitable for all iron contents likely to be found in these classes of materials. If a simple absorptiometer is used the maximum amount that can be determined with sufficient accuracy is about 5% Fe_2O_3; if the amount of ferric oxide exceeds this, the volumetric procedure should be applied.

Colorimetric method

Transfer a 5-ml aliquot of the stock test solution to a 100-ml volumetric flask. Add 2 ml of hydroxyammonium chloride solution (100 g/l), and 5 ml of 1:10-phenantholine solution (10 g/l). Add ammonium acetate solution (approx. 10%) until a pink colour begins to form in the solution, followed by 2 ml in excess.

Allow to stand for 15 min, dilute to 100 ml and mix.

Measure the optical density of the solution against water in 10-mm cells at 510 nm or by using a colour filter (Ilford 603) in a suitable instrument. The colour is stable between 15 min and 75 min after the addition of the ammonium acetate solution.

Determine the ferric oxide content of the solution by reference to a calibration graph.

Using the Unicam SP600 spectrophotometer, the calibration graph extends to about 5% Fe_2O_3. If this amount is exceeded, it is permissible to dilute the stated aliquot to 200 ml in a volumetric flask.

Volumetric method

Take as large an aliquot as possible (usually 100 ml) of the stock test solution. Heat to 80°C, add diluted ammonia solution (1 + 1) until the solution is slightly alkaline to methyl red and filter through a No. 541 Whatman paper. Wash several times with hot water; then wash the bulk of the precipitate back into the precipitation beaker.

Dissolve with hydrochloric acid ($d = 1.18$) any precipitate remaining on the paper and wash thoroughly with hot water into the beaker containing the main precipitate.

Make up the total volume of hydrochloric acid used to 25 ml and warm to dissolve the remainder of the precipitate. Transfer the solution to a conical flask and boil.

While boiling, add stannous chloride solution (30 g/litre), drop by drop, until the solution becomes colourless and then add 1–2 drops in excess.

Cool rapidly to room temperature and add 5 ml of mercuric chloride solution (saturated).

Allow to stand for 10 min, add 15 ml of sulphuric acid-phosphoric acid mixture and 10 drops of barium diphenylamine sulphonate indicator. Titrate to a purple colour with standard potassium dichromate solution (0·05N).

Calculation—1 ml 0·05N $K_2Cr_2O_7$, $\equiv$ 0·00399 g Fe_2O_3.

Determination of Titania

Transfer a 20-ml aliquot of the stock test solution to each of two 50-ml volumetric flasks, A and B. To each flask add 10 ml of diluted phosphoric acid (2 + 3) and to flask A only, 10 ml of hydrogen peroxide solution (6%). Dilute the solution in each flask to 50 ml and shake well.

Measure A against B in 40-mm cells at 398 nm or by using a colour filter (Ilford 601) in a suitable instrument. The colour is stable between 5 min and 24 h after the addition of the hydrogen peroxide solution.

Determine the titania content of the solution by reference to a calibration graph.

Determination of Manganese Oxide

Transfer a 50-ml aliquot of the stock test solution to a 250-ml beaker, add 20 ml of nitric acid ($d = 1·42$) and 10 ml of diluted phosphoric acid (1 + 9).

Boil to remove nitrous fumes, cool slightly and add about 0·2 g of potassium periodate. Boil for 2 min, transfer to a steam bath and keep hot for 10 min.

Allow to cool and transfer to a 100-ml volumetric flask. Dilute the solution in the flask to 100 ml and mix. Measure the optical density of the solution against water in 40-mm cells at 524 nm or by using a colour filter (Ilford 604) in a suitable instrument.

Determine the manganese oxide content of the solution by reference to a calibration graph.

Determination of Alumina

By calculation

Unless results of the highest accuracy are required, the alumina content can be calculated. Subtract the sum of the weights of ferric oxide, titania and manganese oxide from the weight of the mixed oxides obtained from the ammonia precipitation after removal of the residual silica to find the weight of alumina in the sample. If required, phosphate may be determined on a separate sample by any suitable method (e.g. Chapter 21).

By determination

If results of the highest accuracy are required the alumina may be determined as oxinate or by titration with EDTA by the following procedures:

Separation of ferric oxide and titania. Transfer an aliquot (200 ml for high-silica materials, 100 ml for aluminosilicates and aluminous materials) of the stock test solution to a 500-ml separating funnel. Adjust the acidity by adding 20 ml of hydrochloric acid ($d = 1·18$) per 100-ml aliquot. To the aliquot, appropriately acidified, add 20 ml of chloroform and 10 ml of cupferron solution (60 g/litre). Stopper the funnel and shake vigorously. Release the pressure in the funnel by carefully removing the stopper and rinse the stopper and neck of the funnel with water.

Allow the layers to separate and withdraw the chloroform layer. Confirm that extraction is complete by checking that the addition of a few drops of cupferron solution (60 g/litre) to the aqueous solution does not produce a permanent coloured precipitate.

Add further 10-ml portions of chloroform and repeat the extraction until the chloroform layer is colourless. Discard the chloroform extracts.

(a) *Gravimetric determination of alumina.* Transfer the aqueous solution from the cupferron-chloroform separation to a 400-ml beaker. Boil for a few minutes to remove traces of chloroform and then cool.

Neutralize with ammonia solution ($d = 0.88$) to the change point of methyl red and then make just acid with hydrochloric acid ($d = 1.18$). In order to ensure the full recovery of the small amount of alumina normally present in high-silica materials, with these materials add 10 ml of the standard aluminium solution (1.0 mg Al_2O_3/ml) before adjusting the acidity. For work of the highest accuracy, the alumina content of 10 ml of the standard addition should be determined at the same time.

Add 4 ml of diluted hydrochloric acid ($1 + 1$) in excess.

Add 2 ml of hydroxyammonium chloride solution (500 g/litre) and 2 ml of 1:10-phenanthroline solution (10 g/litre) and heat to 40°P50°C. The volume of the solution at this stage should be about 250–300 ml.

Add the requisite amount of 8-hydroxyquinoline solution (50 g/litre) and then add slowly, with stirring, 40 ml of ammonium acetate solution (approx. 40%). For solutions containing up to 30 mg of Al_2O_3, use 10 ml of 8-hydroxyquinoline solution; for solutions containing 30–65 mg of Al_2O_3, use 20 ml of 8-hydroxyquinoline solution: for solutions containing 65–90 mg of Al_2O_3, use 30 ml of 8-hydroxyquinoline solution.

Heat, with stirring, to 60–65°C and allow to stand at this temperature for 10 min stirring periodically. Allow to cool for 30 min.

Filter through a weighed crucible with a sintered-glass mat (porosity 4), scrubbing the beaker with a "bobby". Wash the precipitate with warm water (40–50°C).

Dry the crucible with the precipitate in an oven maintained at 150°C until constant weight is obtained; 2 h is usually sufficient.

For work of the highest accuracy, dissolve the precipitate by passing small quantities of nearly boiling, diluted hydrochloric acid ($1 + 1$) through the crucible and wash thoroughly with hot water. This method of acid extraction is to correct for small quantities of acid-insoluble material occasionally precipitated with the aluminium complex. Dry the crucible and residue at 150°C for 1 h, cool and weigh.

Calculation—Weight of aluminium oxinate $\times$ 0.1110 = weight of Al_2O_3.

(b) *Volumetric determination of alumina.* Run the aqueous solution from the cupferron–chloroform separation into a 500-ml conical flask. Add a few drops of bromophenol blue indicator followed by ammonia solution ($d = 0.88$) until just alkaline.

Re-acidify quickly with hydrochloric acid ($d = 1.18$) and add 5–6 drops in excess.

Add sufficient standard EDTA solution (0·05M) to provide an excess of a few millilitres over the expected amount. Then add ammonium acetate buffer solution until the indicator turns blue, followed by 10 ml in excess. Boil the solution for 10 min and cool rapidly.

Add an equal volume of ethanol (95%) and 1–2 ml of dithizone indicator and titrate with standard zinc solution (0·05M) from blue-green to the first appearance of a permanent pink colour.

Calculation—If the EDTA solution is not exactly 0·05M calculate the equivalent volume of 0·05M EDTA.

1 ml 0·05M EDTA $\equiv$ 2·55 mg Al_2O_3.

Determination of Lime

Adjust the volume of the solution reserved for the determination of lime to about 250 ml by evaporation and add 1 g of oxalic acid.

Boil the solution and add, with stirring, diluted ammonia solution (1 + 1) until the solution is alkaline to bromophenol blue and then add an excess of 10 ml of diluted ammonia solution (1 + 1).

Cover the beaker with a clock glass and digest on a steam bath for 2 h.

Allow to cool and stand for at least 1 h. Filter through a Buchner funnel with a sintered-glass mat (porosity 4) or a No. 42 Whatman paper, and wash the precipitate four times with cold ammonium oxalate solution (10 g/litre).

Just acidify the filtrate and washings with hydrochloric acid ($d = 1.18$) and reserve for the determination of magnesia.

Dissolve the precipitate through the filter with 20 ml of hot, dilute nitric acid (1 + 1) and wash thoroughly with hot water.

Transfer the solution and washings back to the precipitation beaker. Add about 0·2 g of oxalic acid, boil the solution and precipitate the calcium as before.

Cover the beaker with a clock glass and digest on a steam bath for 2 h.

Allow to cool and stand for at least 1 h. Filter through a Buchner funnel with a sintered-glass mat (porosity 4) or a No. 42 Whatman paper and wash the precipitate thoroughly with cold water.

Add the filtrate and washings to the solution reserved for the determination of magnesia and just acidify with hydrochloric acid ($d = 1.18$).

Dissolve the precipitate through the filter with 50 ml of hot, diluted sulphuric acid (1 + 9) and wash the filter thoroughly with hot water.

Transfer the solution and washings back to the precipitation beaker, heat to boiling, and titrate with standard potassium permanganate solution (approx. 0·1N).

Calculation—1 ml 0·1N $KMnO_4$ $\equiv$ 0·002804 g CaO.

Determination of Magnesia

Evaporate the acidified solution reserved for the determination of magnesia to about 300 ml and add 10 ml of ammonium phosphate solution (100 g/litre). Heat to boiling and make just alkaline to bromophenol blue with diluted ammonia solution (1 + 1).

Cool to room temperature and add 20 ml of ammonia solution (d = 0·88). Stir vigorously to start the precipitation and allow to stand overnight at a low temperature.

Filter through a No. 42 Whatman paper and wash four times with cold, diluted ammonia solution (1 + 39), discarding the filtrate and washings.

Dissolve the precipitate through the filter with 20 ml of hot, diluted hydrochloric acid (1 + 1) and wash thoroughly with hot water.

Transfer the solution and washings back to the precipitation beaker. Add 1 ml of ammonium phosphate solution (100 g/litre), heat the solution to boiling, and make just alkaline to bromophenol blue with diluted ammonia solution (1 + 1). Cool, add 10 ml of ammonia solution (d = 0·88), and stir vigorously.

Stand overnight at a low temperature.

Filter through a No. 42 Whatman paper, transferring the precipitate to the paper and scrubbing the beaker with a "bobby". Wash thoroughly with cold, diluted ammonia solution (1 + 39), discarding the filtrate and washings.

Transfer the paper and precipitate to a weighed platinum crucible and heat gently to char the paper and remove carbonaceous matter. (Platinum crucibles may be seriously attacked during the ignition if the precipitate is not thoroughly washed with diluted ammonia solution, or if the precipitate is ignited too strongly before all the carbon is oxidized.) Finally ignite at 1150°C for 15 min, cool and weigh the residue as magnesium pyrophosphate.

Calculation—Weight of $Mg_2P_2O_7 \times 0.3623$ = weight of MgO.

Determination of Alkalis

Decomposition of the Sample

Weigh the finely ground, dried (110°C) sample into a small platinum basin (0·250 g for aluminosilicates and aluminous materials, 1·00 g for high-silica materials) and ignite gently to remove organic matter.

To the cool dish add 10 ml of sulphuric–nitric acid mixture and about 10 ml of hydrofluoric acid (40% w/w).

Transfer the vessel to a sand bath, allow to react thoroughly with the lid on for about 15 min, then evaporate to dryness in a fume cupboard, being careful to avoid spurting.

Cool, add 10 ml of sulphuric–nitric acid mixture, and rinse down the sides of the basin with water. Evaporate carefully to dryness.

To the cool, dry residue add 20 ml of diluted nitric acid (1 + 19) and digest on a steam bath until dissolved.

Cool, filter if necessary, and dilute the solution to 250 ml in a volumetric flask, to form the sample solution A.

Determination of Alkalis

Determine the alkalis in the sample flame-photometrically by the method described in Chapter 16 under "High-silica, Aluminosilicate and Aluminous Materials".

CHAPTER 7

ANALYSIS OF HIGH-SILICA
MATERIALS AND ALUMINOSILICATES:

"RAPID" METHOD

GENERAL

The method described in this chapter is in some sense intermediate between the standard method and the spectrophotometric method. Its accuracy $+ 0$–0.5% for silica and alumina is rather better than the latter, and worse than the standard method; in speed it is slower than spectrophotometry and now, strangely enough, in view of the generally used name, no faster than the routine control modification of the coagulation method. The main justification for its inclusion in the present work is that it still finds a place in a number of laboratories.

It could well continue for some time to be useful, in the smaller laboratory, where the frequency of analysis may not justify either platinum ware or a 1200°C furnace on the one hand, or a spectrophotometer on the other.

The "rapid" method has, like the spectrophotometric method, a restricted scope and is only applicable with the following provisos:

(1) The sample must be capable of decomposition by a sodium hydroxide fusion to ensure a clear solution in hydrochloric acid. This means that materials containing more than about 42% Al_2O_3 cannot be analysed. Samples containing less than this amount of alumina, but to which alumina has been added in an insoluble form, may also be outside the scope of this method.

(2) The sample must also be either soluble in nitric, sulphuric and hydrofluoric acids or be sufficiently attacked to release all the alkalis, lime and magnesia.

(3) The presence of all but traces of phosphate or vanadate in the sample will vitiate the results for silica content.

(4) Wide deviation from a normal silicate or aluminosilicate structure in the material being analysed may give rise to serious errors.

(5) This scheme of analysis is only suitable for materials containing no other elements than are included in the following scheme of analysis.

(6) If much iron is present in the sample ($>5\%$ Fe_2O_3) results for silica and alumina contents can be expected to be below the true values.

(7) The presence of appreciable quantities of lime and particularly magnesia may tend to hold some alumina during the separation of the latter from iron and titanium. This trouble can be overcome by dissolving the residue after this separation, precipitating the R_2O_3 group with ammonia and finally repeating the separation from iron and titanium using sodium hydroxide.

PRINCIPLE OF THE METHOD

The loss on ignition is determined by heating the sample to 1000°C for 30 min. The main part of the analysis is carried out on a portion of the sample fused in sodium hydroxide, the melt decomposed with hot water and dissolved by pouring into hydrochloric acid. The technique ensures that all the silica goes into solution. Aliquots of this solution are used for the determination of silica, alumina, ferric oxide and titania. Silica is determined by first making the aliquot strongly alkaline and then re-acidifying, to ensure that all the silica is in the correct form to react with ammonium molybdate, the reagent used to form the yellow silicomolybdate. The acidity is then increased to prevent the precipitation of quinoline molybdate, and quinoline added to form quinoline silicomolybdate. After heating to ensure the coagulation of the precipitate the solution is allowed to cool, and the precipitate filtered through a sintered-glass crucible, dried and weighed.

The alumina is determined after separation of iron and titanium by a sodium hydroxide precipitation together with the nickel which may have been derived from the crucible. The alumina passes through the filter as aluminate and this is then re-acidified.

Further aliquots of the stock solution are used for the spectrophotometric determinations of ferric oxide with 1:10 phenanthroline after reduction with hydroxyammonium chloride, and titania with hydrogen peroxide after bleaching with phosphoric acid the yellow colour due to ferric salts.

A separate sample solution is prepared by hydrofluoric, nitric and sulphuric acid decomposition of the sample, sulphuric and nitric acids are used to remove fluoride by evaporation to dryness and the residue is dissolved in diluted nitric acid and made up to volume. The alkalis in the solution are determined flame-photometrically and aliquots are used for the compleximetric determinations of lime and lime plus magnesia, the ammonia group oxides being complexed, in each case, with triethanolamine. Lime is determined by adjusting the pH with potassium hydroxide and titrating with EDTA using screened Calcein as indicator, and lime plus magnesia by adjusting the pH with ammonia solution and titrating with EDTA using methylthymol blue complexone as indicator.

REAGENTS

Additional Reagents

Quinoline: boiling range 235–239°C
Thymol blue

Prepared Reagents

Ammonium acetate (approx. 40%): Dilute 570 ml acetic acid (glacial) to 1700 ml with water and add carefully 570 ml of ammonia solution ($d = 0.88$) mix, cool and adjust to pH 6·4–6·8, either with acetic acid or ammonia solution.

Ammonium acetate solution (approx. 10%): Dilute the ammonium acetate (approx. 40%) with three times its volume of water.

Ammonium molybdate (100 g/litre): Dissolve 100 g of ammonium molybdate in 1 litre of water; filter if necessary. Store in a polythene bottle and discard after 4 weeks or earlier if any appreciable deposit of molybdic acid is observed.

8-Hydroxyquinoline (50 g/litre): Dissolve, by warming, 50 g of 8-hydroxy-quinoline in 120 ml of acetic acid (glacial). Dilute to about 700 ml, filter, cool and dilute to 1 litre.

1:10 phenanthroline (10 g/litre): Prepare enough solution for immediate use at a concentration of 0·1 g of 1:10 phenanthroline hydrate in 10 ml of diluted acetic acid (1 + 1).

Quinoline (2% v/v): Add 20 ml of quinoline (boiling range 235°C–239°C) to about 800 ml of hot water acidified with 25 ml of hydrochloric acid ($d = 1·18$) stirring constantly if cloudy. Cool, add some paper pulp and stir vigorously. Allow to settle and filter under suction through a paper-pulp pad to remove traces of oily matter, but do not wash. Dilute the filtrate to 1 litre.

Quinoline wash solution (0·05% v/v): Dilute 25 ml of the quinoline solution (2% v/v) to 1 litre.

Sulphuric-nitric acid mixture: To 650 ml of water, add 100 ml of diluted sulphuric acid (1 + 1) and 250 ml of nitric acid ($d = 1·42$).

Indicators
    Calcein (screened)
    Methyl red
    Methylthymol blue complexone
    Thymol blue

Standard Solution
    EDTA (5 g/litre)

PREPARATION OF SAMPLE

The sample prepared for analysis should be ground to pass completely through a 120-mesh B.S. test sieve. A non-metallic (e.g. nylon bolting-cloth) sieve is preferable.

High-silica materials and aluminosilicates containing up to 42% alumina may generally be ground in agate, or in iron with subsequent treatment with a magnet.

PROCEDURE

Determination of Loss on Ignition

Weigh 1·000 g of the finely ground, dried (110°C) sample into a platinum crucible. Heat the crucible over a low mushroom flame, slowly increasing the temperature to full heat over a period of about 20 min. Transfer the crucible to a muffle furnace at 1000°C and ignite to constant weight; 30 min is usually sufficient.

Determination of Silica, Alumina, Ferric Oxide and Titania

The weight of sample taken depends on the anticipated silica content: not more than 40 mg of silica should be present in the aliquot taken for the silica determination. A much larger sample weight than is needed is therefore taken and a suitable aliquot of the fusion extract is used for the determination.

(a) For samples containing less than 65% of silica, weigh 0·500 g of the material. Dilute the fusion solution to 250 ml as indicated in the procedure and take an aliquot calculated to yield 25–35 mg of silica. For example, if the sample contains 50% of silica, a 25-ml aliquot will contain 25 mg.

(b) For samples containing more than about 65% of silica, weigh 0·250 g of sample. Dilute the fusion solution to 250 ml and take an aliquot calculated to contain 25–35 mg of silica.

Decomposition of the Sample

Fusion

Carefully fuse 7 g of sodium hydroxide pellets in a nickel crucible (40-mm dia. and 40-mm deep) until the water is driven off and a clear melt is obtained. Allow to cool and brush the weighed sample on to the cold melt.

Gently tap the crucible to spread the sample evenly over the surface of the sodium hydroxide. Moisten the sample with ethanol (95%) and carefully evaporate the alcohol on a hot plate. Provided reasonable care is taken, this will prevent the light sample being blown from the crucible during the fusion.

Gently fuse over a mushroom burner, occasionally swirling the crucible, until the sample is dissolved and the melt is quiet, then increase the temperature to a dull red heat; 5 min is usually sufficient for the fusion.

Extraction of the fused mass

Carefully cool the mass by running water round the outside of the crucible until the melt just solidifies. Place the still hot crucible in a 250-ml nickel beaker, and cover with a clock glass. Raise the cover slightly, fill the crucible with water at the boiling point and quickly replace the cover glass. The crucible should be hot enough to boil the water and dissolve the fused mass. When the vigorous reaction is over, wash the cover glass and the sides of the beaker with hot water and remove the crucible with a pair of tongs, carefully rinsing it inside and out with hot water and scrubbing it with a "bobby".

Measure 20 ml of hydrochloric acid ($d = 1·18$) into a 500-ml conical flask and, after adjusting the volume of the fusion extract to about 175 ml, pour into the acid, swirling continuously during the addition. Complete the transfer with a jet of hot water, scrubbing out the beaker with a "bobby".

The flask may be gently heated, if necessary, until the solution clears, but do not boil. Cool rapidly to room temperature and dilute to 250 ml in a volumetric flask.

This "stock" solution is used for the determination of silica, alumina, titania and sometimes ferric oxide.

Determination of Silica

Once the fusion has been started, the procedure should be continued at least to the end of the precipitation, otherwise the results may not be accurate. The precipitate may, if necessary, be stood in the cold overnight. Bright sunlight should not be allowed to fall on the precipitate while it is standing, otherwise a small amount of decomposition may occur.

Transfer the required aliquot of the stock solution to a 400-ml polythene beaker without diluting, add 3 g of sodium hydroxide pellets, swirling until dissolved, and dilute to about 250 ml.

Add 10 drops of thymol blue indicator followed by hydrochloric acid ($d = 1\cdot18$) dropwise, swirling continuously, until the indicator changes from blue, through yellow, just to red. The solution should become clear almost immediately.

Transfer the solution to an 800-ml beaker, add 10 ml of diluted hydrochloric acid ($1 + 9$) and dilute to about 400 ml.

Add 50 ml of ammonium molybdate solution (100 g/litre) from a pipette or burette, stirring vigorously during the addition and for 1 min after it, and stand at room temperature for 10 min.

Add 50 ml of hydrochloric acid ($d = 1\cdot18$) and immediately precipitate the quinoline silicomolybdate by adding 50 ml of quinoline solution (2% v/v) from a burette, stirring the solution during the addition.

Heat to 80°C over a period of about 10 min and maintain this temperature for 5 min. Cool in running water to less than 20°C and filter through a weighed crucible with a sintered-glass mat (porosity 4) transferring the precipitate to the crucible with a jet of quinoline wash solution, also cooled to less than 20°C. After scrubbing out the beaker with a "bobby", wash the precipitate eight times with quinoline wash solution, care being taken never to let the precipitate run dry during filtration and washing.

Dry the precipitate at 150°C to constant weight; 2 h is usually sufficient. Cool in a desiccator and weigh. The precipitate may be dried at 110–120°C for a minimum of 2 h, but in this case results may be fractionally high.

Calculation—Weight of precipitate $\times$ $0\cdot02566$ = weight of SiO_2.

Determination of Alumina

Place 5 g of sodium hydroxide pellets into a 250-ml nickel beaker and then transfer a 50-ml aliquot of the stock solution to the beaker. Boil for 5 min.

Allow the precipitate to settle, then filter the solution through a No. 541 Whatman paper into 400-ml beaker containing 20 ml of diluted hydrochloric acid ($1 + 1$), transferring the precipitate to the filter with a jet of hot sodium carbonate solution (10 g/litre); wash. The filtrate should now be slightly alkaline.

Neutralize the filtrate with diluted hydrochloric acid ($1 + 1$) to the change point of methyl red indicator and add 4 ml of diluted hydrochloric acid ($1 + 1$) in excess. Boil for 1 min to remove most of the carbon dioxide.

The volume should now be about 200 ml. Add 4 ml of hydroxyammonium chloride solution (500 g/litre) and 5 ml of 1:10 phenanthroline solution (10 g/litre), and heat the solution to 40–50°C.

Add 20 ml of 8-hydroxyquinoline solution (50 g/litre), then add slowly, with stirring, 40 ml of ammonium acetate solution (approx. 40%) Heat the solution, with stirring, to 60–65°C and allow to stand at this temperature for 10 min, stirring periodically, and then allow to cool for 30 min.

Filter through a weighed crucible with a sintered-glass mat (porosity 4) transferring the precipitate to the filter with a jet of warm water (40–50°C), wash and then dry the crucible and precipitate in an air oven at 150°C to constant weight; 2 h is usually sufficient.

If it is suspected that any silica has precipitated with the alumina complex (indicated by a tendency to filter slowly), dissolve the weighed precipitate through the filter with small quantities of nearly boiling, diluted hydrochloric acid $(1 + 1)$ and wash the filter thoroughly with hot water. Dry the crucible for 1 h at 150°C, cool and weigh.

Calculation—Weight of aluminium oxinate $\times$ 0·1110 = weight of Al_2O_3.

Determination of Ferric Oxide

The determination of ferric oxide can be carried out on either the solution prepared by the sodium hydroxide fusion or the solution prepared by hydrofluoric acid attack. As it has been proved that there is a danger of the loss of iron into nickel crucibles the second method is preferable, provided that a clear solution has been obtained.

Transfer a 5-ml aliquot of the solution to a 100-ml volumetric flask. Add 2 ml of hydroxyammonium chloride (100 g/litre), 5 ml of 1:10-phenanthroline solution (10 g/litre) and ammonium acetate solution (approx 10%), drop by drop, until a faint pink colour develops, followed by 2 ml in excess.

Allow to stand for 15 min, dilute to 100 ml and mix.

Measure the optical density of the solution against water in 10-mm cells at 510 nm or by using a colour filter (Ilford 603) in a suitable instrument. The colour is stable between 15 min and 75 min after the addition of the ammonium acetate solution.

Determine the ferric oxide content of the solution by reference to a calibration graph.

Determination of Titania

Transfer a 40-ml aliquot of the stock solution to each of two 100-ml volumetric flasks, A and B. To each flask add 10 ml of diluted phosphoric acid $(2 + 3)$ and, to flask A only, 10 ml of hydrogen peroxide solution (6%).

Dilute the solution in each flask to 100 ml and shake well.

Measure A against B in 40-mm cells at 398 nm or by using a colour filter (Ilford 601) in a suitable instrument. The colour is stable between 5 min and 24 h after the addition of the hydrogen peroxide solution.

Determine the titania content of the solution by reference to a calibration graph.

Determination of Lime, Magnesia and Alkalis

Decomposition of the Sample

Weigh 0·250 g of the finely ground, dried (110°C) sample into a platinum basin and ignite gently to remove organic matter.

To the cool dish add 10 ml of sulphuric–nitric acid mixture and about 10 ml of hydrofluoric acid (40% w/w). Transfer the vessel to a sand bath in a fume cupboard, allow to react thoroughly with the lid on for about 15 min, then evaporate to dryness, being careful to avoid spurting.

Cool, add 10 ml of the sulphuric–nitric acid mixture and rinse down the sides of the basin with water. Evaporate carefully to dryness.

To the cool dry residue add 20 ml of diluted nitric acid (1 + 19) and warm to dissolve.

Cool, filter if necessary and dilute the solution to 250 ml in a volumetric flask, to form the sample solution A.

Determination of Lime

Transfer a 100-ml aliquot of the sample solution A, prepared by the attack of hydrofluoric acid, to a 500-ml conical flask. Add 5 ml of diluted triethanol-amine (1 + 1) and 10 ml of potassium hydroxide solution (250 g/litre) and dilute to about 200 ml.

Add about 0·015 g of screened Calcein indicator and titrate with standard EDTA solution (5 g/litre) from a semi-micro burette, the colour change being from fluorescent green to pink.

Determination of the Sum of Lime and Magnesia

Transfer a 100-ml aliquot of the sample solution A, prepared by the attack of hydrofluoric acid, to a 500-ml conical flask. Add 10 drops of hydrochloric acid ($d = 1·18$), 20 ml of diluted triethanolamine (1 + 1) and 25 ml of ammonia solution ($d = 0·88$). Dilute to about 200 ml.

Add about 0·04 g of methylthymol blue complexone indicator and titrate with standard EDTA solution (5 g/litre) from a semi-micro burette, the colour change being from blue to colourless.

Calculation of Magnesia

The volume of EDTA used for the titration of lime is subtracted from the volume of EDTA used for the titration of the sum of lime and magnesia. The remainder represents the volume of EDTA required for the titration of the magnesia.

Determination of Alkalis

Determine the alkalis in the sample flame-photometrically by the method described in Chapter 16 under "High-silica and Aluminosilicate Materials".

CHAPTER 8

ANALYSIS OF HIGH-SILICA MATERIALS AND ALUMINOSILICATES:

SPECTROPHOTOMETRIC METHOD

GENERAL

The method described in this chapter is intended solely for routine control purposes. In view of the speed of the coagulation method, particularly when used as the routine control modification (Chapter 5), the spectrophotometric method is mainly useful when results for silica and alumina *must* be obtained in about 2 h. Where full analyses are required, there is little gain in speed over the coagulation technique. In addition, the latter has the advantage of greater reliability in results, for although errors for silica and alumina by the spectrophotometric method are normally less than $\pm 0.5\%$, experience has shown that an occasional sport result, without apparent reason, is possible.

However, even though the main value of the method now lies in the very rapid determination of silica and alumina, it is still in fairly widespread use and full details are therefore given. The method is applicable with the following provisos:

(1) The sample must be capable of decomposition by a sodium hydroxide fusion to ensure a clear solution in nitric acid. This means that materials containing more than about 42% Al_2O_3 cannot be analysed. Samples containing less than this amount of alumina, but to which alumina has been added in an insoluble form may also be outside the scope of the method.

(2) The sample must also be either soluble in nitric, sulphuric and hydrofluoric acids or be sufficiently attacked to release all the alkalis, lime and magnesia.

(3) The presence of all but traces of phosphate or vanadate in the sample will vitiate the results for silica content.

(4) Wide deviation from a normal silicate or aluminosilicate structure in the materials being analysed may give rise to serious errors.

(5) This scheme of analysis is only suitable for materials containing no other elements than are included in the following methods.

PRINCIPLE OF THE METHOD

The loss on ignition is determined by heating the sample to 1000°C for 30 min. The main part of the analysis is carried out on a portion of the sample fused in sodium hydroxide, the melt extracted with hot water, dissolved in nitric acid and the solution made up to volume. Aliquots are taken for silica, alumina and titania determinations by spectrophotometric procedures. Silica is determined by measuring the yellow silicomolybdate colour and alumina by the addition of Solochrome cyanine, measuring against a control solution.

The iron in one aliquot is complexed with thioglycollic acid, and the iron and aluminium in the control with EDTA. Titanium is determined with hydrogen peroxide, using phosphoric acid to bleach the yellow colour of the iron. Ferric oxide is not normally determined on this solution as there is a tendency to lose some of the iron in the sample into the nickel crucible.

A separate sample solution is prepared by hydrofluoric, nitric and sulphuric acid decomposition of the sample, nitric and sulphuric acids are used to remove fluoride by evaporation to dryness and the residue dissolved in diluted nitric acid and made up to volume. The alkalis in the solution are determined flame-photometrically and the ferric oxide in an aliquot is determined spectrophotometrically with 1:10 phenanthroline after reduction with hydroxyammonium chloride. Further aliquots are used for the compleximetric determinations of lime, and lime plus magnesia, the ammonia group oxides being complexed, in each case, with triethanolamine. Lime is determined by adjusting the pH with potassium hydroxide and titrating with EDTA using screened Calcein as indicator, and lime plus magnesia is determined by adjusting the pH with ammonia solution and titrating with EDTA using methylthymol blue complexone as indicator.

REAGENTS

Additional Reagents

Solochrome cyanine
Thioglycollic acid

Prepared Reagents

Ammonium acetate (approx. 40%): Dilute 570 ml of acetic acid (glacial) to 1700 ml with water and add carefully 570 ml of ammonia solution ($d = 0.88$). Mix, cool and adjust the pH 6.0 (using a pH meter) either with acetic acid or ammonia solution. Make enough of this solution to last as long as the Solochrome cyanine solution. (It has been found that a change of ammonium acetate solution can have a marked effect on the calibration).

Ammonium acetate (approx. 10%): Dilute the ammonium acetate solution (approx. 40%) with three times its volume of water.

Ammonium molybdate (80 g/litre): Dissolve 80 g of ammonium molybdate in 1 litre of water; filter if necessary. Store in a polythene bottle and discard after 4 weeks or earlier if any appreciable deposit of molybdic acid is observed.

1:10 Phenanthroline (10 g/litre): Prepare enough solution for immediate use at a concentration of 0.1 g of 1:10 phenanthroline hydrate in 10 ml of diluted acetic acid (1 + 1).

Solochrome cyanine (0.75 g/litre): Dissolve 0.75 g of Solochrome cyanine in 200 ml of water. Add 25 g of sodium chloride, 25 g of ammonium nitrate and 2 ml of nitric acid ($d = 1.42$). Dilute to 1 litre with water. Filter through a No. 42 Whatman paper but do not wash. Stand for 24 h before use. This solution will deteriorate after 4 weeks.

Sulphuric acid–phosphoric acid mixture: Cautiously add 200 ml of sulphuric acid ($d = 1.84$) and 500 ml of phosphoric acid ($d = 1.75$) to 300 ml of water,

E

cooling the solution and keeping it well mixed during the addition of the acids.

Sulphuric–nitric acid mixture: To 650 ml of water, add 100 ml of diluted sulphuric acid $(1 + 1)$ and 250 ml of nitric acid $(d = 1.42)$.

Indicators

Calcein (screened)
Methylthymol blue complexone

Standard Solution

EDTA (5 g/litre)

PREPARATION OF SAMPLE

The sample for analysis should be ground to pass completely through a 120-mesh B.S. test sieve. A non-metallic (e.g. nylon bolting-cloth) sieve is preferable.

High-silica materials and aluminosilicates containing up to 42% Al_2O_3 may generally be ground in agate, or in iron with subsequent treatment with a magnet.

PROCEDURE

Determination of Loss on Ignition

Weigh 1·000 g of the finely ground, dried (110°C) sample into a platinum crucible. Heat the crucible over a low mushroom flame, slowly increasing the temperature to full heat over a period of about 20 min. Transfer the crucible to a muffle furnace at 1000°C and ignite to constant weight: 30 min ignition is usually sufficient.

Determination of Silica, Alumina and Titania

On unfired materials containing a large amount of carbonaceous matter it is often advantageous to use the sample on which the loss on ignition has been determined for the fusion, otherwise some carbon may be left in the final solution.

Once the fusion has been started, the determinations should be completed forthwith, otherwise the results may not be accurate.

Fusion

Gently fuse 3 g of sodium hydroxide pellets in a nickel crucible (40-mm dia., 40-mm deep) until any water is driven off and a clear melt is obtained. Allow to cool and brush exactly 0·2000 g of the finely ground, dried (110°C), or ignited, sample on to the solidified melt.

Lightly tap the crucible to spread the sample evenly over the surface of the sodium hydroxide, moisten the sample with about 1 ml of ethanol (95%) and gently evaporate the alcohol on a hot plate. Provided that reasonable care is taken, this will prevent the light sample being blown from the crucible during the early stages of the fusion.

Gently fuse over a burner, occasionally swirling the contents of the crucible until the sample is dissolved and the melt is quiet, then increase the temperature to a dull red heat; 5 min is usually sufficient for the fusion.

Extraction of the Fused Mass

Carefully cool the mass by running cold water round the outside of the crucible until the melt just solidifies. Place the still hot crucible in a 250-ml nickel beaker and cover with a clock glass.

Raise the cover slightly, fill the crucible with water at the boiling point and quickly replace the cover glass. The crucible should be hot enough to boil the water and dissolve the fused mass. When the vigorous reaction is over, wash the cover glass and the sides of the beaker with hot water and remove the crucible, carefully rinsing it inside and out with hot water and scrubbing it out with a "bobby".

Measure out and transfer 40 ml of diluted nitric acid $(1 + 2)$ into a 500-ml conical beaker or flask and, after adjusting the volume of the fusion extract to about 175 ml, pour it into the acid, swirling constantly during the addition. Complete the transfer with a jet of hot water, scrubbing out the beaker with a "bobby".

The solution must be cleared by heating, if necessary, but it must not be boiled. Cool rapidly to room temperature, dilute to 500 ml in a volumetric flask and mix to give the stock solution.

Determination of Silica

Transfer a 25-ml aliquot of the stock solution to each of two 100-ml volumetric flasks, A and B. To each add 10 ml of diluted nitric acid $(1 + 4)$ and dilute to about 50 ml with water.

To flask A only, add 40 ml of ammonium molybdate solution (80 g/litre) with swirling, noting the exact time at which the molybdate enters the solution. Dilute the solution in each flask to 100 ml and shake well.

Measure A against B in 10-mm cells at 440 nm, not earlier than 5 min and not later than 15 min after the addition of the ammonium molybdate solution. Five replicate readings should be taken and the mean reading calculated. Determine the silica content of the solution by calculation from a calibration equation.

Determination of Alumina

Special care must be taken to avoid contamination by hydrofluoric acid fumes, because even as little as 0·1 mg F can cause negative errors of 3 % Al_2O_3 at the 30 % level.

Samples containing 10–40 % alumina

Transfer a 20-ml aliquot of the stock solution to a 500-ml volumetric flask, dilute to 500 ml and mix. Transfer a 20-ml aliquot of this diluted solution to each of two 250-ml volumetric flasks, A and B, and add sufficient water to each to bring the volume to about 100 ml.

To flask A add 10 ml of diluted thioglycollic acid (1 + 199), and to flask B, 5 ml of EDTA solution (5 g/litre). Then add to each flask 25 ml of Solochrome cyanine solution (0·75 g/litre) and 20 ml of ammonium acetate solution (approx. 40%). Dilute the solution in each flask to 250 ml and shake well.

Stand for 10 min at room temperature and measure A against B in 10-mm cells, at 536 nm, rinsing the cells with the solution at least six times before the final filling.

Five replicate readings should be taken and the mean reading calculated. Determine the alumina content of the solution by reference to a calibration graph.

Samples containing 0–2% alumina

Transfer a 20-ml aliquot of the stock solution to each of two 250-ml volumetric flasks, A and B, and dilute to about 100 ml.

To flask A add 10 ml of diluted thioglycollic acid (1 + 199) and to flask B, 5 ml of EDTA solution (5 g/litre). Then add to each flask 25 ml of Solochrome cyanine solution (0·75 g/litre) and 20 ml of ammonium acetate solution (approx. 40%).

Dilute the solution in each flask to 250 ml, and shake well. Stand for 10 min at room temperature, and measure A against B in 10-mm cells at 536 nm.

Five replicate readings should be taken and the mean reading calculated. Determine the alumina content of the solution by reference to a calibration graph.

Determination of Titania

Transfer an 80-ml aliquot of the stock solution to each of two 100-ml volumetric flasks, A and B. To the solution in each flask, add 5 ml of sulphuric acid-phosphoric acid mixture and to flask A only, 10 ml of hydrogen peroxide solution (6%).

Dilute the solution in each flask to 100 ml and shake well.

Measure A against B in 40-mm cells at 398 nm.

Determine the titania content of the solution by reference to a calibration graph.

Determination of Ferric Oxide, Lime, Magnesia and Alkalis

Decomposition of the Sample

Weigh 0·250 g of the finely ground sample, previously dried at 110°C into a small platinum basin and ignite gently to remove organic matter. To the cool basin add 10 ml of sulphuric-nitric acid mixture, 10 ml of hydrofluoric acid (40% w/w) and evaporate to dryness on a sand bath in a fume cupboard, being careful to avoid spurting. Cool, add 10 ml of the sulphuric-nitric acid mixture and rinse down the sides of the dish with water. Evaporate carefully to dryness.

To the cool, dry residue add 20 ml of dilute nitric acid (1 + 19) and warm to dissolve. Cool, filter if necessary through a No. 40 Whatman paper and wash with cold water. Dilute the filtrate and washings, with water, to 250 ml in a volumetric flask to form the sample solution A.

Determination of Ferric Oxide

The ferric oxide is best determined on the solution prepared by the attack of hydrofluoric acid, but if this part of the analysis is not required, it is possible to determine the ferric oxide by using a portion of the sodium hydroxide fusion solution. In this case the calibration graph should be constructed by fusion of standard samples of known ferric oxide content, but the results will be less accurate.

Transfer a 5-ml aliquot of the sample solution A from the hydrofluoric acid decomposition to a 100-ml volumetric flask. Add 2 ml of hydroxyammonium chloride solution (100 g/litre), 5 ml of 1:10-phenanthroline solution (10 g/litre) and ammonium acetate solution (approx. 10%) until a pink colour appears in the solution, then add 2 ml in excess.

Allow to stand for 15 min, dilute to 100 ml and mix.

Measure the optical density of the solution against water in 10-mm cells at 510 nm.

Determine the ferric oxide content of the solution by reference to a calibration graph.

Determination of Lime

Transfer a 100-ml aliquot of the sample solution A prepared by the hydrofluoric acid decomposition to a 500-ml conical flask. Add 5 ml of diluted triethanolamine (1 + 1) and 10 ml of potassium hydroxide solution (250 g/litre). Dilute to about 200 ml.

Add about 0·015 g of screened Calcein indicator and titrate with standard EDTA solution (5 g/litre) from a semi-micro burette, the colour change being from fluorescent green to pink. The end point is the final change in colour.

Determination of the Sum of Lime and Magnesia

Transfer a 100-ml aliquot of the sample solution A prepared by the hydrofluoric acid decomposition to a 500-ml conical flask and add 10 drops of hydrochloric acid ($d = 1·18$). Add 20 ml of diluted triethanolamine (1 + 1) and 25 ml of ammonia solution ($d = 0·88$). Dilute to about 200 ml.

Add about 0·04 g of methylthymol blue complexone indicator and titrate with standard EDTA solution (5 g/litre) from a semi-micro burette, the colour change being from blue to colourless.

Calculation of Magnesia

The volume of EDTA used in the titration of lime is subtracted from the volume of EDTA used in the titration of the sum of lime and magnesia. The remainder represents the volume of EDTA required for the titration of the magnesia.

Determination of Alkalis

Determine the alkalis in the sample flame-photometrically by the method described in Chapter 16 under "High-silica and Aluminosilicate Materials".

CALIBRATION OF THE INSTRUMENT

Silica

Reagents

Additional reagents

Standard samples of known silica content may be used as an alternative to silica. (British Chemical Standards 267, 269, 313, 314 and 315 are recommended and are available from the Bureau of Analysed Samples Ltd., Newham Hall, Middlesbrough, Yorks.)

Procedure

Standard determination

Carry out fusions either on silica of known purity or on standard samples of known silica content, and determine the optical density scale reading. With pure silica, fusions should be carried out with 0·2 and 0·1 g, giving points at approximately the 100% and 50% levels respectively. Alternatively, standard samples at about the 96% and 50% levels are recommended. The calibration samples should be treated exactly as for an ordinary determination.

Since the calibration forms the basis for calculation of a considerable number of determinations, it is essential to make its accuracy as high as possible. It is suggested therefore that four fusions be carried out for each of the two levels of silica contents and that determinations of the optical density should be made from two aliquots from each solution, making a total of eight mean readings. The mean of these eight is then accepted.

Calculation of the calibration

The appropriate cell blanks should be deducted from each of the eight readings for each level of silica and the corrected values averaged to give two gross means.

x_1 being the gross mean for the higher silica content, $y_1\%$
and x_2 being the gross mean for the lower silica content, $y_2\%$

It has been found that the graph of optical density plotted against silica content is a straight line which corresponds to the general formula:

$$y = mx - c$$

where　y is the silica content (percentage)
　　　　m is the slope of the line
　　　　x is the optical density
　　　　c is a constant.

For the general case, the slope of the line, m, is obtained first, as follows:

$$m = \frac{y_1 - y_2}{x_1 - x_2}$$

The constant, c, is due to the silica present in the reagents; it thus constitutes the "reagent blank" and is expressed therefore as a percentage of silica. This can be obtained as follows:

$$\text{blank} (\%) = c = m \, (2x_2 - x_1) + y_1 - 2y_2$$

Alumina

Reagents

Standard solutions

Standard aluminium: $1 \cdot 0$ mg Al_2O_3/ml.

Intermediate aluminium standard: Dilute 20 ml of the standard aluminium solution to 250 ml in a volumetric flask: 1 ml $\equiv 0 \cdot 08$ mg Al_2O_3.

Dilute aluminium standard (10–40% range): Dilute 20 ml of the intermediate aluminium standard solution to 500 ml in a volumetric flask. Do not keep longer than 24 h: 1 ml $\equiv 0 \cdot 0032$ mg Al_2O_3.

Dilute aluminium standard (0–2% range): Dilute 25 ml of the intermediate aluminium standard solution to 500 ml in a volumetric flask. Do not keep longer than 24 h: 1 ml $\equiv 0 \cdot 004$ mg Al_2O_3.

Blank solutions
Blank fusion solution (10–40% range): Carry out a fusion, as detailed on p. 82, omitting the sample. Dilute the extract to 500 ml with water.

Blank fusion solution (0–2% range): Carry out a fusion, as detailed on p. 82, on 0·2 g of pure silica. Dilute the extract to 500 ml with water.

Procedure

10–40% range

With the concentration of alumina chosen in the dilute standard, 1 ml $\equiv$ 1% Al_2O_3. Therefore the method is calibrated at 20% and 40% by the use of 20 ml and 40 ml of dilute standard. To each of two 250-ml volumetric flasks, A and B, add 20 ml of dilute blank fusion solution (10–40% range) and the required volume of dilute aluminium standard (10–40% range). Dilute to approximately 100 ml.

To flask A add: 10 ml of diluted thioglycollic acid (5 g/litre), 25 ml of Solochrome cyanine solution (0·75 g/litre) and 20 ml of ammonium acetate solution (approx. 40%).

To flask B add: 5 ml of EDTA solution (5 g/litre), 25 ml of Solochrome cyanine solution (0·75 g/litre) and 20 ml of ammonium acetate solution (approx. 40%).

Make each flask up to the mark with water, shake well and stand for 10 min at room temperature. Measure A against B (i.e. setting Check on B) at

536 nm in 10-mm cells, rinsing the cells at least six times with the solution before finally filling.

Calculate the mean density readings after deduction of cell blanks and find the calibration equation as below.

An adequate number of determinations should be made to ensure the accuracy of the two calibration points.

0-2% range

With the concentration of alumina chosen in the dilute standard, 20 ml $\equiv$ 1% Al_2O_3. Therefore the method is calibrated at 0·5% and 2·0% by the use of 10 ml and 40 ml of the dilute standard. Proceed as follows:

To each of two 250-ml volumetric flasks, A and B, add 20 ml of the blank fusion solution (0–2% range) and the required volume of dilute aluminium standard (0–2% range). Dilute to approximately 100 ml.

To flask A add: 10 ml of diluted thioglycollic acid (5 g/litre), 25 ml of Solochrome cyanine solution (0·75 g/litre) and 20 ml of ammonium acetate solution (approx. 40%).

To flask B add: 5 ml of EDTA solution (5 g/litre), 25 ml of Solochrome cyanine solution (0·75 g/litre) and 20 ml of ammonium acetate solution (approx. 40%).

Make each flask up to the mark with water, shake well and stand for 10 min at room temperature. Measure A against B (i.e. setting Check on B) at 536 nm in 10-mm cells, rinsing the cells at least six times with the solution before finally filling.

Calculate the mean density readings after deduction of cell blanks and find the calibration equation as below.

Calculation

Samples containing 10–40% Al_2O_3

If y_1 is the mean density reading for 40%,
y_2 the mean density reading for 20%,
y the mean density reading for the sample,
then the percentage alumina, x, is given by:

$$x = 20 + \frac{20(y - y_2)}{y_1 - y_2}.$$

Samples containing 0·00–2·00% Al_2O_3

If y_1 is the mean density reading for 2·0%,
y_2 the mean density reading for 0·5%,
y the mean density reading for the sample,
then the percentage alumina, x, is given by:

$$x = 0·5 + \frac{1·5(y - y_2)}{y_1 - y_2}.$$

The above equations are derived from the general formula

$$x = m(y - c)$$

where m is the slope of the straight line given by plotting the alumina content against the density reading and c is a constant due to the reagent blanks.

Titania

Reagents

Standard solutions

Standard titanium: 0·64 mg TiO_2/ml
Dilute titanium standard: Dilute 50 ml of the standard titanium solution to 500 ml in a volumetric flask:

1 ml $\equiv$ 0·064mg TiO_2 (0·064mg TiO_2 $\equiv$ 0·20% TiO_2 in the sample)

Blank solution: Double strength—carry out a fusion, as detailed on p. 82 omitting the sample but using double quantities of sodium hydroxide and nitric acid.

Procedure

With the concentration of titanium chosen in the dilute standard, 25 ml $\equiv$ 5% TiO_2. Therefore, the method is calibrated at 0, 1, 3 and 5% TiO_2 by the addition of 0, 5, 15 and 25 ml of the dilute standard to 40-ml aliquots of the double-strength blank solution. Proceed as follows: to each of two 100-ml volumetric flasks, A and B, transfer 40 ml of double-strength blank solution and the required volume of dilute titanium standard. To flask A add: 5 ml of sulphuric acid–phosphoric acid mixture and 10 ml of hydrogen peroxide (6%). To flask B add: 5 ml of sulphuric acid–phosphoric acid mixture.
Make each flask up to the mark with water, shake well and measure A against B in 40-mm cells at 398 nm.
Calibrate each point in duplicate and plot a graph of density against percentage TiO_2.

Ferric Oxide

NOTE: If the ferric oxide content of the sample is determined on an aliquot of the sodium hydroxide fusion solution, the calibration curve should be constructed by fusion of standard samples of known ferric oxide content. British Chemical Standards 267, 269, 314 and 315 are suitable.

Reagents

Standard solutions

Standard iron: 0·1 mg Fe_2O_3/ml.
Dilute iron standard: Dilute 100 ml of the standard iron solution to 1 litre in a volumetric flask: 1 ml $\equiv$ 0·01 mg Fe_2O_3.
Blank Solution: Dilute 20 ml of diluted nitric acid (1 + 19) to 250 ml.

E*

Procedure

With the concentration of iron chosen in the dilute standard, 1 ml $\equiv 0.2\%$ Fe_2O_3. Therefore, to 5-ml portions of blank solution add 0, 5, 25 and 50 ml of dilute standard iron solution and develop the colour as detailed on p. 85. Check each point in duplicate and plot a graph of density against percentage Fe_2O_3.

NOTES ESSENTIAL FOR ACCURATE OPERATION OF THE METHOD

Weighing

As only 0.2000 g is used as a sample the errors caused by inaccuracies in the weighing become very important, e.g. an error of 0.0010 g can be equivalent to a final error of 0.5% SiO_2 or 0.2% Al_2O_3. Hence not only must the balance be accurate and in good condition, but the weighing must be completed rapidly to prevent undue moisture absorption.

Transference of Sample

For the above reasons extreme care is necessary, when brushing the sample on to the sodium hydroxide and when adding the alcohol, to avoid loss of the fines in the sample.

Decomposition of the Sample

The melting of the sodium hydroxide must be carried out slowly to prevent mechanical loss of sample by spurting. It is desirable to achieve a compromise in the fusion time, so that the sample is completely decomposed without introduction of an excessive quantity of nickel.

Withdrawal of Aliquots

In order to eliminate one source of pipette error it is desirable to use the same pipette for sample solutions and for calibration. Cleanness of the pipette itself and utmost accuracy in its use are vital, as an error of 0.1 ml can be equivalent to an error of 0.4% SiO_2 or 0.2% Al_2O_3. All pipettes should be cleaned regularly with chromic acid solution.

Colour Development

The reagents should always be added in the correct order and into a similar volume of solution.

Careful note should be made of the time when the ammonium molybdate is first added for silica, and the ammonium acetate for alumina and iron.

A white precipitate forms as the molybdate is added, but this normally clears in a few seconds. If it remains after the solution has been made up to volume and mixed, the aliquot should be rejected.

Care of Cells

Utmost cleanliness is essential. Daily cleaning with chromic acid followed by liberal leaching with distilled water is normally adequate.

The clear optical faces should not be handled and, after the cells have been filled, should be wiped dry with a lintless cloth or paper tissue.

The optical faces should be inspected after insertion in the cell carriage to check that they are clean and dry.

"Cell blanks" should be checked at least daily, any appreciable deviation from normal indicating lack of cleanness.

Filling of Cells

Each cell should be rinsed out at least six times with the solution to be used in it. After completion of the readings the cells should be emptied and rinsed similarly with distilled water.

Standing Time for Silica

The colour should be measured not earlier than 5 min after the first addition of molybdate, because the yellow colour is then incompletely developed; and not later than 15 min after, as then the colour begins to fade. Between 5 and 15 min the fading is normally less than would be equivalent to 0.1% SiO_2.

Operation of the Instrument

It should be pointed out that the method achieves its accuracy only by using the spectrophotometer at the utmost limit of the precision that can reasonably be expected of it. It is therefore necessary to perform every operation involved in its use to the finest point of detail, particularly with reference to the slit width and optical density controls. Further modifications of the method may possibly assist in this but their success cannot yet be assumed.

Wavelength Setting

The nature of the absorption band of the yellow silicomolybdate complex is such that it is impossible to work on a flat portion of the curve. For this reason extremely precise setting of the wavelength drum at 440 nm is of the greatest importance. The errors introduced by slight inaccuracies are of considerable magnitude and must therefore be minimized by utmost care in the setting. If this is done, errors not greater than 0.1% would be expected when the wavelength is reset.

In this connection it is advisable to use the left hand for operating the cell carriage to avoid accidental movement of the wavelength control.

Slit Width

This control is operated to obtain a zero reading on the blank solution with the instrument set at "Check". Again extreme care is necessary here to set the pointer exactly on the zero line, particularly as the control is very sensitive; a fractional movement of the knob causes a considerable movement of the pointer and thus results in a large potential error.

Alignment of the Cells

It is advisable to use each cell in one and the same compartment of the carrier every time, to eliminate any possibility of positional errors.

Considerable errors can be introduced by failing to align the cell carrier into the cell carriage and also by failing to confirm that the carriage drops accurately into its appropriate notch.

Accuracy of Readings

It is possible to read the optical density scale very accurately at the lower end, but when the silica content approaches 100% and the optical density reading about 0·800, individual divisions become very small. The present method results in 1 scale division of the optical density scale representing about $1·25\%$ SiO_2 and $0·6\%$ Al_2O_3, so it is obviously desirable to attempt to estimate the reading to 0·001, even though this is very difficult. It is doubtful whether errors much greater than 0·002 can be tolerated. For this reason five separate readings should be taken and the mean reading used for the calculation of the percentage silica content.

Calibration of the Instrument

As the calibration takes almost a whole day, it should not be undertaken more frequently than necessary. It is recommended that a new calibration should be prepared when fresh solutions of ammonium molybdate, Solochrome cyanine, and ammonium acetate are made up. Other reagents should therefore be made up to last as long as these reagents will remain usable—normally about 4 weeks, after which time they should be discarded. Before use, the condition of the molybdate solution should be checked, and if there is any appreciable indication of white flakes at the bottom of the bottle it is advisable to discard the reagent. This is particularly true when determinations are being made in the region of 100% silica content, as the method will not then tolerate any appreciable reduction in the molybdate concentration.

For silica it is necessary to calculate the slope of the curve to four significant figures, because at the 100% level a change of 1 in the fourth figure affects the result by almost $0·1\%$.

Duplicate Determinations

To ensure that accuracies of the order of $\pm0·5\%$ are achieved, two fusions should be carried out and a determination made on each of the two resulting solutions. Normally these duplicates will agree well and any major discrepancy between them (e.g. $> 0·5\%$) must give rise to doubts as to the value of the mean.

CHAPTER 9

DIRECT DETERMINATION OF ALUMINA IN HIGH-SILICA MATERIALS

The alumina content of silica rocks and high-silica refractory products is vital, and often needs to be determined without carrying out any other part of the analysis. For this reason, direct methods for the determination of alumina in these materials have been devised, one of which is a British Standard method.

Although the latter is the oxine gravimetric method, the EDTA volumetric method is currently preferred. This incorporates a cleaner separation of iron and titanium from alumina, by means of solvent extraction, and also avoids the undesirable addition of alumina to the solution to ensure complete precipitation. This addition has become increasingly frequent because of the greater purity of silica rocks used for the manufacture of silica bricks.

Both the current British Standard method, and the more recent and preferred volumetric method are given in this chapter.

PREPARATION OF SAMPLE

The sample prepared for analysis should be ground to pass completely through a 120-mesh B.S. test sieve. A non-metallic (e.g. nylon bolting-cloth) sieve is preferable.

The bulk sample may be crushed in an iron mortar or by a jaw crusher and, after coning and quartering, the iron is removed with a magnet. The final grinding is done best in an agate mortar.

A. Oxine Gravimetric Method
(British Standard B.S. 1902: Part 2A: 1964)

PRINCIPLE OF THE METHOD

The sample is decomposed with hydrofluoric and sulphuric acids; the weight used depends on the amount of alumina present. After re-evaporation to dryness with more sulphuric acid, traces of fluoride and sulphate are removed by ignition to 1000°C. The residue is fused in sodium carbonate/boric acid, and the melt decomposed with water. Insoluble material is filtered off; thus removing iron and titanium and much of the alkaline earths, and the filtrate and washings acidified with hydrochloric acid.

For accurate work, the residue is re-dissolved in hydrochloric acid and the iron and titanium reprecipitated in strongly alkaline conditions with sodium hydroxide. The solution is again filtered, when any traces of alumina pass through as aluminate, and the two filtrates are combined.

Hydroxyammonium chloride and 1:10 phenanthroline are added to complex traces of iron escaping the precipitations, oxine is added and the alumina precipitated by buffering the pH of the solution. After digestion and cooling, the precipitate is filtered on a sintered-glass Gooch crucible, dried and weighed.

When the alumina content is low, there is a threshold value below which incomplete recovery is achieved, so that alumina must be added from a standard solution. In order to achieve maximum accuracy when this is done, it is desirable to precipitate and recover the alumina content of the amount of standard solution at the same time. Blank determinations are normally necessary as they often reach significant proportions.

RANGE

From 0·1 % to 10% alumina.

PRECISION

With 1 g of sample and 1 % of alumina present: $\pm$ 0·03 % of alumina.

REAGENTS

Prepared Reagents

Ammonium acetate (approx. 40%): Dilute 570 ml of acetic acid (glacial) to 1700 ml with water and add carefully 570 ml of ammonia solution ($d = 0\cdot88$). Mix, cool and adjust to pH 6·4–6·8, either with acetic acid or ammonia solution.

8-Hydroxyquinoline (50 g/litre): Dissolve, by warming, 50 g of 8-hydroxyquinoline in 120 ml of acetic acid (glacial). Dilute to about 700 ml, filter, cool and dilute to 1 litre.

1:10 Phenanthroline (10 g/litre): Prepare enough solution for immediate use at a concentration of 0·1 g of 1:10 phenanthroline hydrate in 10 ml of diluted acetic acid (1 + 1).

Indicator

Methyl red

Standard Solution

Aluminium (1·0 mg Al_2O_3/ml)

PROCEDURE

For materials containing 0·1–0·5% of alumina use 2 g of sample, for materials containing 0·5–5% use 1 g of sample and for materials containing 5–10% use 0·5 g of sample.

Accurately weigh the appropriate amount of the finely ground, dried (110°C) sample into a platinum dish or crucible; a convenient size is 50-mm dia. and 45-mm deep.

Moisten with water, add 5 drops of sulphuric acid ($d = 1\cdot84$) followed by 10 ml of hydrofluoric acid (40 % w/w). Transfer the vessel to a sand bath in a fume cupboard, digest for 15 min and then evaporate to dryness; cool, moisten the residue with water, add 5 drops of sulphuric acid ($d = 1\cdot84$) and evaporate to dryness. Heat cautiously over a mushroom burner until fumes of sulphur trioxide cease to be evolved and then raise the temperature to 1000°C and ignite for 5 min.

Thoroughly mix the residue with 2·5 g of anhydrous sodium carbonate and 1 g of boric acid, heat at a moderate temperature until the salts fuse, and then complete the fusion at 1000°C for 20 min.

Allow the contents of the crucible to cool and add distilled water to the fused mass. Heat the contents of the covered crucible almost to boiling over a very low flame until the mass is disintegrated.

Filter the contents of the crucible through a No. 40 Whatman paper, collecting the filtrate in a 400-ml beaker. Rinse the crucible several times with hot sodium carbonate solution (10 g/litre) and pour the washings through the filter paper. Finally, wash the paper thoroughly with hot sodium carbonate solution (10 g/litre) and just acidify the filtrate with diluted hydrochloric acid (1 + 1) to the change point of methyl red.

For work of the highest accuracy any alumina left in the residue from the fusion should be recovered as follows:

Rinse the platinum crucible with the minimum amount of diluted hydrochloric acid (1 + 1) and pour the rinsings and washings through the filter paper. Add hot, diluted hydrochloric acid (1 + 1) to the filter paper, drop by drop, in such a way that all parts of the paper, particularly the edges, are moistened with the acid. Collect the filtrate in a 100-ml beaker and wash the paper with hot water several times.

The residue from the fusion should be completely soluble and no gritty particles should be left.

Nearly neutralize the filtrate and washings with sodium hydroxide solution (100 g/litre) and then pour the solution into an equal volume of sodium hydroxide solution (100 g/litre) contained in a vessel of platinum or nickel— but not of porcelain or glass.

Boil for 5 min, add a small amount of macerated filter paper and filter through the paper previously used. Wash this paper several times with hot sodium carbonate solution (10 g/litre).

Make the filtrate and washings just acid with diluted hydrochloric acid (1 + 1) and add them to the 400-ml beaker containing the main portion of the alumina.

Add 10 ml of diluted hydrochloric acid (1 + 9) and boil for 1 min to remove carbon dioxide. Cool to room temperature.

Add 4 ml of hydroxyammonium chloride (500 g/litre) and 5 ml of 1:10-phenanthroline solution (10 g/litre) and heat to 40–50°C. Add the requisite amount of 8-hydroxyquinoline solution (50 g/litre). For solutions containing up to 30 mg of alumina add 10 ml of 8-hydroxyquinoline solution, for solutions containing 30–65 mg of alumina add 20 ml of 8-hydroxyquinoline solution.

Add slowly, with stirring, 40 ml of ammonium acetate solution (approx. 40%), heat, with stirring, to 60–65°C and allow to stand at this temperature for 10 min, stirring periodically.

Filter through a weighed crucible with a sintered-glass mat (porosity 4) and wash the precipitate thoroughly with warm water at 40–50°C.

Dry the crucible and contents in an oven maintained at 150°C until constant weight is obtained; 2 h is usually sufficient.

For work of the highest accuracy, dissolve the precipitate by passing small quantities of nearly boiling, diluted hydrochloric acid $(1 + 1)$ through the crucible and wash thoroughly with hot water. Dry the crucible at 150°C for 1 h, cool and weigh.

Calculation—Weight of aluminium oxinate $\times$ 0·1110 = weight of Al_2O_3.

B. EDTA Volumetric Method

PRINCIPLE OF THE METHOD

An appropriate sample is taken, based on the anticipated alumina content, and decomposed with hydrofluoric and sulphuric acids, followed by evaporation to dryness. A further evaporation with sulphuric acid followed by ignition is carried out to ensure the removal of fluorides.

The residue is fused in sodium carbonate and boric acid and the melt dissolved in hydrochloric and sulphuric acids. Iron and titanium are removed by means of a cupferron/chloroform extraction and acidity of the solution adjusted. Excess EDTA is added, the solution buffered, boiled and cooled. An equal volume of alcohol is added, followed by dithizone indicator and the excess EDTA back-titrated with zinc solution.

RANGE

The method is designed for materials with alumina contents of $<2\%$ but. by suitable adjustment of the sample weight and/or the volume of EDTA added, the method is satisfactory for alumina contents of up to about 5%.

PRECISION

With 1% of alumina, S.D. $= 0·02\%$

REAGENTS

Prepared Reagents

Ammonium acetate buffer: Add 120 ml of acetic acid (glacial) to 500 ml of water followed by 74 ml of ammonia solution $(d = 0·88)$. Mix, cool and dilute to 1 litre.

Cupferron (60 g/litre): Dissolve 3 g of cupferron in 50 ml of cold water and filter. This solution must be freshly prepared.

If the reagent is discoloured or gives a strongly coloured solution a new stock should be obtained. The solid reagent should be stored in a tightly-stoppered bottle in the presence of a piece of ammonium carbonate to prevent decomposition.

Indicators
  Bromophenol blue
  Dithizone

Standard Solutions

EDTA (0·05M)

Zinc (0·05M)

DECOMPOSITION OF THE SAMPLE

Weigh 1·000 g of the finely ground, dried (110°C) sample into a platinum basin or crucible; a convenient size is 50-mm dia. and 45-mm deep. If it is anticipated that the material contains less than 0·25% alumina, take a 2-g sample. Moisten the sample with water, add 5 drops of sulphuric acid ($d = 1·84$) and 8 ml of hydrofluoric acid (40% w/w), or 15 ml with a 2-g sample.

Cover the vessel with a lid and digest the mixture for 15 min on a hot sand bath in a fume cupboard, to facilitate the attack of the sample before evaporation; then remove the lid and evaporate until fuming freely. Cool, add 2 ml of hydrofluoric acid (40% w/w), or 8 ml with a 2-g sample, and evaporate to dryness. Heat the dry residue cautiously until fumes of sulphur trioxide cease to be evolved and then raise the temperature to 1000°C and ignite for 5 min.

Mix the residue with 2·5 g of anhydrous sodium carbonate and 1 g of boric acid. Heat at a moderate temperature until the salts fuse and then complete the fusion at 1000°C for 20 min, with intermittent swirling of the crucible and its contents throughout this period. Allow the crucible and contents to cool and then transfer them to a 250-ml beaker provided with a cover glass. Add 30 ml of water, 10 ml of hydrochloric acid ($d = 1·18$) and 25 ml of diluted sulphuric acid (1 + 1). When dissolution of the melt is complete remove the crucible and lid and wash them thoroughly with hot water.

DETERMINATION OF ALUMINA

Cool and transfer the solution to a 500-ml separating funnel. Add 20 ml of chloroform and 10 ml of cupferron solution (60 g/litre). Stopper the funnel and shake vigorously. Release the pressure in the funnel by carefully removing the stopper and rinse the stopper and neck of the funnel with water. Allow the layers to separate and withdraw the chloroform layer. Confirm that extraction is complete by checking that the addition of a few drops of cupferron solution (60 g/litre) does not produce a permanent coloured precipitate. Add further 10-ml portions of chloroform and repeat the extraction until the chloroform layer is colourless. Wash the stem of the funnel, inside and out, with chloroform, using a polythene wash bottle.

Discard the chloroform extracts and transfer the aqueous solution and washings from the funnel to a 500-ml conical flask. Add a few drops of bromophenol blue indicator and then add ammonia solution ($d = 0·88$) until the solution is just alkaline. Re-acidify quickly with hydrochloric acid ($d = 1·18$) and add 5–6 drops in excess. Add sufficient EDTA (0·05M) to provide an excess of a few millilitres over the expected amount. On a 1-g sample, 1 ml of EDTA (0·05M) $\equiv$ 0·255% Al_2O_3. Then add ammonium acetate buffer solution until the indicator turns blue, followed by 10 ml in excess. Boil the solution for 10 min and cool rapidly.

Add an equal volume of ethanol (95%) (if sulphates are precipitated by the alcohol, just sufficient water should be added to redissolve them) followed by 1–2 ml of dithizone indicator. If the solution shows a pinkish tinge it is advantageous to screen out the colour by the dropwise addition of Naphthol Green B solution (1 g/litre). Titrate with standard zinc solution (0·05M) from green to the first appearance of a permanent pink colour.

Calculation—If the EDTA solution is not exactly 0·05M, calculate the equivalent volume of exactly 0·05M EDTA.

If W is the weight of the sample in grammes,
    V ml is the volume of EDTA (0·05M) and
    v ml is the volume of zinc solution used in back-titration,
    then the percentage of alumina, x, is given by:

$$x = \frac{0·255\,(V - v)}{W}$$

CHAPTER 10

DIRECT DETERMINATION OF ALUMINA IN ALUMINOSILICATES AND ALUMINOUS MATERIALS

GENERAL

A knowledge of the alumina content of fireclays, firebricks and aluminous raw materials and products is of considerable commercial importance. For this reason a direct method for this determination has been devised and has been issued as a British Standard method.

Where only the alumina content of the material is required, it is convenient to use the method described in this chapter, even though the accuracy of the method is probably less than can be achieved by using the single-dehydration method. The British Standard for the direct method quotes an accuracy of $\pm 0.3\%$ Al_2O_3.

A more accurate procedure, but a more lengthy one, is to apply the method described in Chapter 5, removing the silica by coagulation, re-fusing the silica residue and determining the alumina content with EDTA. Techniques such as fusion, and dissolution, without separation of silica, or HF treatment and fusion of the residue, followed by a cupferron separation and the application of the EDTA volumetric procedure, are commonly used for routine purposes. These methods, although quite reproducible do yield results which appear to be about $\frac{1}{4}\%$ low; this is often considered adequate in view of the saving in time.

PREPARATION OF THE SAMPLE

The sample for analysis should be ground to pass completely through a 120-mesh B.S. test sieve. A non-metallic (e.g. nylon bolting-cloth) sieve is preferable.

Aluminosilicates containing up to 45% of alumina may be ground in agate, or in iron with subsequent treatment with a magnet.

Aluminous materials may be ground in boron carbide, alumina or iron mortars; in the latter case the ground sample should be treated with a magnet. For work of the highest accuracy, it may be necessary to prepare two samples and carry out the alumina determination on the iron-ground sample and, by determining the iron content of both samples, make allowance for the amount of iron introduced as contamination.

PRINCIPLE OF THE METHOD

The sample is fused in sodium carbonate and boric acid and the melt dissolved in hydrochloric acid. Iron and titanium are removed by making the solution strongly alkaline with sodium hydroxide and boiling. The sodium aluminate, in solution, is filtered and the filtrate just reacidified. Oxine is then added and the solution buffered. After digesting and cooling, the precipitate is filtered on a sintered-glass Gooch crucible, dried and weighed. As some traces of silica etc. may also be retained, for accurate work the precipitate is redissolved and the Gooch crucible washed thoroughly and re-weighed.

Low results tend to occur if the bulk of the alkali precipitate is large; high ferric oxide contents and particularly high magnesia contents tending to retain alumina. These errors can partially be overcome, if the ferric oxide content is high; redissolving the precipitate in acid and reprecipitating is effective, but at the same time, the procedure increases the amount of iron passing into the solution. Large amounts of magnesia entails redissolving the alkali precipitate, removing the magnesia by an ammonia precipitation, redissolving and again precipitating with sodium hydroxide. This multiplicity of separation greatly reduces the effectiveness of the procedure.

Blank determinations on the reagents are essential.

REAGENTS

Prepared Reagents

Ammonium acetate (approx. 40%): Dilute 570 ml of acetic acid (glacial) to 1700 ml with water and add carefully 570 ml of ammonia solution ($d = 0.88$). Mix, cool and adjust to pH 6·4–6·8, either with acetic acid or ammonia solution.

8-Hydroxyquinoline (50 g/litre): Dissolve, by warming, 50 g of 8-hydroxy-quinoline in 120 ml of acetic acid (glacial). Dilute to about 700 ml, filter, cool and dilute to 1 litre.

1:10 Phenanthroline (10 g/litre): Prepare enough solution for immediate use at a concentration of 0·1 g of 1:10 phenanthroline hydrate in 10 ml of diluted acetic acid (1 + 1).

Indicator

Methyl red

Standard Solution

Aluminium (1·0 mg Al_2O_3/ml)

PROCEDURE

Accurately weigh 0·1 g of the finely ground, dried (110°C) sample into a platinum dish or crucible; weigh in 0·25 g of anhydrous sodium carbonate and 0·08 g of boric acid. Mix well to form a charge about 20 mm dia. in the middle of the vessel and cover with a lid.

Heat cautiously at first, gradually raise the temperature to about 900°C during a period of about 10 min. Finally heat for about 15 min at 1200°C in a muffle furnace.

Remove from the furnace and allow to cool.

Add 5 ml of diluted hydrochloric acid (1 + 1) and allow the dish or crucible to stand on a steam bath with occasional stirring until the sintered mass is completely decomposed; 20–30 min is normally sufficient.

Transfer the contents of the platinum vessel to a 100-ml beaker and nearly neutralize with sodium hydroxide solution (100 g/litre) keeping the volume to about 30 ml. Pour the solution slowly, with stirring, into 30 ml of sodium hydroxide solution (100 g/litre) contained in a platinum or nickel vessel, adding the rinsings from the beaker. Heat to boiling and boil for about 5 min.

Allow the precipitate to settle. Filter through a No. 541 Whatman paper into about 15 ml of hydrochloric acid ($d = 1\cdot18$). Wash the precipitate thoroughly with hot sodium hydroxide solution (20 g/litre) and finally with hot water.

Discard the paper and residue and add a few drops of methyl red indicator followed by diluted ammonia solution (1 + 1) until the indicator just changes to yellow; just re-acidify, then add 4 ml of diluted hydrochloric acid (1 + 1) in excess.

Add 4 ml of hydroxyammonium chloride solution (500 g/litre) and 5 ml of 1:10-phenanthroline solution (10 g/litre) and heat to 40–50°C. Add the requisite amount of 8-hydroxyquinoline solution (50 g/litre); use 20 ml of 8-hydroxyquinoline solution for samples containing up to 65% of alumina, or 30 ml of 8-hydroxyquinoline solution for samples containing more than 65% of alumina.

Add slowly, with stirring, 40 ml of ammonium acetate solution (approx. 40%). Heat the solution, with stirring, to 70°C and allow to stand at this temperature for 10 min, stirring periodically. Allow it to cool for 30 min.

Filter through a weighed crucible with a sintered-glass mat (porosity 4). and wash the precipitate thoroughly with warm water (40–50°C).

Dry the crucible and precipitate in an air oven at 150°C until constant weight is obtained; 2 h is usually sufficient.

For work of the highest accuracy, dissolve the precipitate by passing small quantities of nearly boiling, diluted hydrochloric acid (1 + 1) through the crucible and wash thoroughly with hot water. Dry the crucible at 150°C for 1 h, cool and weigh. This method of acid extraction is to correct for small amounts of acid-insoluble material which may be precipitated with the aluminium complex.

Calculation—Weight of aluminium oxinate $\times$ $0\cdot1110$ = weight of Al_2O_3.

Blank Determination

Corrections for "blank" are essential and the determination of the blank should be carried out as is described in the procedure, the sample being omitted. To avoid undue attack on the platinum the flux should be dissolved directly in the acid and not fused. In order to ensure the full recovery of the small amount of aluminium in the reagents, 5 ml of the standard aluminium solution (1·0 mg Al_2O_3/ml) should be added immediately before the 8-hydroxyquinoline precipitation, and the aluminium content of this amount of aluminium solution ascertained at the same time. The blank is determined by the difference between the amount of alumina found and that added.

CHAPTER 11

ANALYSIS OF MAGNESITES AND DOLOMITES

GENERAL

The second edition of "Methods of Silicate Analysis" by Bennett and Hawley described two methods for the analysis of magnesites and dolomites, the "classical" and a routine control method. At that time, both were needed because the routine method was not completely reliable. Since then this method has been extensively investigated on a co-operative basis and has been radically modified. It has recently been submitted to the British Standards Institution and accepted as a standard method. Thus, there is no longer any need to include the classical method as an alternative.

PREPARATION OF SAMPLE

The sample prepared for analysis should be ground to pass completely through a 120-mesh B.S. test sieve. A non-metallic (e.g. nylon bolting-cloth) sieve is preferable.

Many of these materials are comparatively soft and may be ground in agate mortars without appreciable contamination. In a few cases the material may be sufficiently hard to be contaminated and then two samples must be prepared, one ground in iron and treated with a magnet and the other ground in agate. The first sample is used for the main analysis and the second for the iron determination, the results obtained on the first sample being corrected for the iron contamination.

PRINCIPLE OF THE METHOD

With the exception of raw dolomites, which are stable under atmospheric conditions, it is normal to refer analyses of magnesites and dolomites to the ignited basis. Progressive hydration and carbonation take place on exposure and only the ignited basis can provide a satisfactory form for comparison. It is desirable, therefore, to weigh all the requisite portions of sample for analysis at the same time as those used for the determination of loss on ignition.

Determination of Loss on Ignition:

The loss on ignition is determined at $1025 \pm 25°C$.

Determination of Silica, Ferric Oxide, Titania, Manganese Oxide, Chromic Oxide, Alumina, Lime and Magnesia

The sample is decomposed with hydrochloric acid, and the silica separated by coagulation and filtration. The crucible containing this silica is ignited and weighed before and after treatment with hydrofluoric and sulphuric acids. The silica remaining in the filtrate is subsequently determined by a spectrophotometric method, based on the formation of molybdenum blue, in a portion of the solution used for the determination of ferric oxide, titania manganese oxide, chromic oxide, alumina, lime and magnesia.

The residue from the silica is fused in sodium carbonate/boric acid and the cold melt dissolved in the filtrate from the silica. After dilution to standard volume, the ferric oxide is determined spectrophotometrically with 1:10-phenanthroline, the titania with hydrogen peroxide, the manganese oxide with potassium periodate and the chromic oxide with diphenylcarbazide for very low chrome contents and as the chrome–EDTA complex for materials containing more than about 0·1% of chromic oxide. For the alumina determination, the ammonia group oxides in a measured volume of the solution are precipitated with ammonia, to remove most of the lime and magnesia. The precipitate is dissolved in hydrochloric acid and the ferric and titanium ions removed by a cupferron/chloroform solvent extraction. The determination is completed volumetrically with diaminoethanetetra-acetic acid (EDTA) and zinc, using dithizone as indicator. Lime is determined on a measured volume of the solution to which triethanolamine is added to complex interfering elements. A known volume of standard EGTA is then added; this is sufficient to complex all the calcium, and the magnesium is then precipitated with potassium hydroxide and filtered off. The excess EGTA in an aliquot of the filtrate is titrated with standard calcium solution, Calcein being used as indicator. Magnesia is determined volumetrically on a measured volume of the solution, after removal of ferric, manganese and most of the titanium ions by a sodium diethyldithiocarbamate/chloroform solvent extraction. Alumina and any remaining titanium ions are complexed with triethanolamine and the magnesium is titrated in a strongly ammoniacal solution containing ammonium chloride, with DCTA. If more than 1% Al_2O_3 is present, this and titania are removed by a buffered cupferron/chloroform solvent extraction. The titration also includes the titration for lime which must be allowed for.

Determination of Alkalis

The sample is decomposed by hydrofluoric, nitric and sulphuric acids to remove the silica. A further evaporation with nitric and sulphuric acids removes traces of fluoride; the residue is dissolved in nitric acid and the solution diluted to a standard volume. Alkalis are determined using a flame photometer.

REAGENTS

Additional Reagents

  Ammonium ceric nitrate
  Diphenylcarbazide
  Magflok
  Sodium azide

Prepared Reagents

Ammonium acetate (approx 10%): Dilute 140 ml of acetic acid (glacial) to 2000 ml with water and add carefully, with stirring, 140 ml of ammonia solution ($d = 0.88$). Mix, cool and adjust to pH 6·0–6·5, either with acetic acid or ammonia solution.

Ammonium acetate buffer (for Al₂O₃ determination): Add, with stirring, 120 ml of acetic acid (glacial) to 500 ml of water, followed by 74 ml of ammonia solution ($d = 0.88$). Cool, dilute to 1 litre and mix.

Ammonium acetate buffer (for MgO determination): Dilute 5 ml of ammonia solution ($d = 0.88$) to 100 ml with distilled water and add 30 ml of acetic acid (glacial). Adjust to pH 3·8 using a pH meter and dilute to 200 ml with distilled water.

Ammonium ceric nitrate (10 g/litre): Dissolve 2·5 g of ammonium ceric nitrate in about 200 ml of water, cautiously add 7 ml of sulphuric acid ($d = 1.84$), cool, dilute to 250 ml and mix.

Ammonium molybdate (80 g/litre): Dissolve 80 g of ammonium molybdate in water, filter if necessary, dilute to 1 litre and mix. Store in a polythene bottle. Discard after 4 weeks, or earlier if any appreciable deposit is observed.

Ammoniacal ammonium nitrate (10 g/litre): Dilute 10 ml of nitric acid ($d = 1.42$) to about 200 ml. Add ammonia solution (1 + 1) until the solution is faintly alkaline to bromophenol blue. Cool and dilute to 1 litre.

Cupferron (60 g/litre): Dissolve 1·5 g of cupferron in 25 ml of water; filter if necessary. This solution must be freshly prepared.

If the reagent is discoloured or gives a strongly coloured solution a new stock should be obtained. The solid reagent should be stored in a tightly-stoppered bottle in the presence of a piece of ammonium carbonate to prevent decomposition.

Diaminoethanetetra-acetic acid, disodium salt, dihydrate (EDTA) (50 g/litre): Dissolve 25 g of the salt in warm water, filter if necessary, cool, dilute to 500 ml and mix. Store in a polythene bottle.

Diphenylcarbazide (10 g/litre): Dissolve 0·1 g of diphenylcarbazide in 10 ml of acetone. This solution must be freshly prepared.

Magflok (20 g/litre approx): The resin is a thick viscous liquid and it is therefore most convenient to transfer a drop to a beaker and to weigh the amount taken. Sufficient water is then added to make up a solution approx. 20 g/litre.

This reagent is available from Ridsdale & Co. Ltd., Newham Hall, Middlesbrough. The material is a partially hydrolysed polyacrylamide.

1:10 Phenanthroline (10 g/litre): Prepare enough solution for immediate use at a concentration of 0·1 g of 1:10 phenanthroline hydrate in 10 ml of acetic acid (1 + 1).

Polyethylene oxide (2·5 g/litre): Add 0·5 g of polyethylene oxide* to 200 ml of water slowly with stirring, preferably on a mechanical stirrer, until dissolved. Discard after 2 weeks.

Sodium diethyldithiocarbamate (100 g/litre): Dissolve 5 g of sodium diethyl-

---

* Union Carbide Polyox resins WSR 35, WSR-N-80, WSR 205, WSR-N-750 and WSR-N-3000 are suitable as a source of polyethylene oxide.

dithiocarbamate in 50 ml of water and filter. This solution must be freshly prepared.

Sodium sulphite (50 g/litre): Dissolve 2 g of sodium sulphite in 40 ml of water. This solution must be freshly prepared.

Stannous chloride (10 g/litre): Dissolve, by warming, 1 g of stannous chloride in 1·5 ml of hydrochloric acid ($d = 1·18$). Cool and dilute to 100 ml. This solution should not be kept more than 24 h.

Sulphurous acid: Saturate 250 ml of water with sulphur dioxide.

Indicators

Bromophenol blue;
Calcein (screened)
2,4-dinitrophenol
Dithizone
Solochrome Black 6B

Standard Solutions

Calcium (0·05M)
DCTA (0·05M)
EDTA (0·05M)
EGTA (0·05M)
Zinc (0·05M)

BLANK DETERMINATIONS

A blank determination shall be carried out on all reagents in accordance with the general scheme of analysis. When carrying out the blank determination for lime, it is necessary to add 50 ml of magnesium sulphate heptahydrate solution (6 g/litre) before addition of the standard EGTA solution.

PROCEDURE

Determination of Loss on Ignition

Weigh 1·000 g of the finely ground sample, previously dried at 110°C, into a platinum crucible. Almost completely cover the crucible with a lid, place in a muffle furnace and slowly raise the temperature to 1025 ± 25°C. Ignite to constant weight at this temperature. Ignition for 30 min is usually sufficient. Remove the crucible from the furnace, completely cover with the lid and weigh as soon as possible.

For routine work, it is possible to start the ignition over a low mushroom flame, slowly increasing the temperature to full heat over a period of about 20 min, after which the crucible is transferred to a furnace at 1000°C for 30 min.

Determination of Silica, Ferric Oxide, Titania, Manganese Oxide, Chromic Oxide, Alumina, Lime and Magnesia

Decomposition of the sample

Weigh 5·000 g of the finely ground sample, previously dried at 110°C, and transfer it to a 250-ml beaker. Add 25 ml of water and 40 ml of hydrochloric acid ($d = 1\cdot18$) and cover with a clock glass. Transfer to a sand bath and boil for 30 min.

Determination of the main silica

Allow the beaker and contents to cool and rinse the clock glass with water into the solution. Add a Whatman accelerator tablet and stir to break up the pulp, then add with stirring, 5 ml of polyethylene oxide solution ($2\cdot5$ g/litre) and allow to stand for 5 min. Filter through a 110-mm fine filter paper (Whatman No. 42 is suitable), and transfer the silica to the filter with hot, diluted hydrochloric acid (1 + 19), scrubbing the beaker with a rubber-tipped glass rod. Wash the precipitate six times with hot, diluted hydrochloric acid (1 + 19) and then with hot water until free from chlorides. Reserve the filtrate and washings.

Transfer the paper and precipitate to an ignited and weighed platinum crucible. Ignite at a low temperature until the precipitate is free from carbonaceous matter and then heat in a muffle furnace at 1200°C to constant weight, 15 min being normally sufficient.

Moisten the contents of the cold crucible with water, add 5 drops of diluted sulphuric acid (1 + 1) and about 10 ml of hydrofluoric acid (40% w/w). Evaporate to dryness on a sand bath in a fume cupboard. For the evaporation, the crucible and contents should be heated from below; the use of top heating alone, as with a radiant heater, can result in incomplete elimination of silica by the hydrofluoric acid.

Heat the crucible and residue, cautiously at first, over a gas flame and finally for 5 min at 1200°C, cool, and weigh. If the residue weighs more than 30 mg, repeat the treatment with sulphuric and hydrofluoric acids to ensure that all the silica is removed. The difference between the two weights represents the "gravimetric" silica.

Preparation of Solution for the Determination of Residual Silica, Ferric Oxide, Titania, Manganese Oxide, Chromic Oxide, Alumina, Lime and Magnesia

Fuse the residue from the hydrofluoric acid treatment of the "gravimetric" silica with 2 g of sodium carbonate and $0\cdot4$ g of boric acid and dissolve the melt in the main filtrate. Cool, dilute the solution to 500 ml in a volumetric flask and mix.

This solution is referred to as the "stock solution".

Determination of Residual Silica

Transfer $5\cdot0$ ml of the stock solution to a 100-ml volumetric flask, A, and add 15 ml of water. Add 2 drops of 2:4-dinitrophenol indicator and diluted ammonia solution (1 + 1) dropwise until the indicator turns yellow (note the amount of ammonia used), then add 5 ml of diluted hydrochloric acid (1 + 4).

To another 100-ml volumetric flask, B, add 20 ml of water and the same amount of diluted ammonia solution (1 + 1) as used to neutralize the aliquot in flask A. Add 2 drops of 2:4-dinitrophenol indicator followed by diluted hydrochloric acid (1 + 4) until the solution is neutral and then 5 ml in excess.

To both flasks add 6 ml of ammonium molybdate (80 g/litre) and stand for

5 to 10 min at a temperature of not less than 20°C and not greater than 30°C. Then add, with swirling, 45 ml of diluted hydrochloric acid (1 + 1) and stand for 10 min. Add 10 ml of stannous chloride solution (10 g/litre), dilute to 100 ml and mix. (The deep yellow-brown colour which appears on the addition of the stannous chloride is quite normal and does not interfere at the wavelength used). Measure the optical density of the solution in flask A against the solution in flask B in 10-mm cells at 800nm, or by using a colour filter (Ilford 609) in a suitable instrument.

The colour is stable between 5 min and 30 min after the addition of the stannous chloride solution.

Determine the silica content of the solution by reference to a calibration graph.

Calculation of Total Silica Content

Add the figure for residual silica content to that obtained for the "gravimetric" silica to obtain the total silica content.

Determination of Ferric Oxide

This determination is for total iron expressed as ferric oxide. Any iron normally present in the sample should have been oxidized to the ferric state during the determination of loss on ignition. Thus the recording of total iron as ferric oxide takes account of this and results in true analysis totals.

Dilute 50·0 ml of the stock solution to 250 ml in a volumetric flask and mix. This solution is referred to as the "diluted stock solution" and is also used for the determination of magnesia.

Transfer 5·00 ml of the "diluted stock solution" to a 100-ml volumetric flask. Add 2 ml of hydroxyammonium chloride solution (100 g/litre), 5 ml of 1:10 phenanthroline solution (10 g/litre) and 2 ml of ammonium acetate solution (approx. 10%). Allow to stand for 15 min, dilute to 100 ml and mix.

Measure the optical density of the solution against water in 10-mm cells at 510 nm, or by using a colour filter (Ilford 603) in a suitable instrument. The colour is stable for 15 min to 75 min after the addition of the ammonium acetate solution. Determine the ferric oxide content of the solution by reference to a calibration graph.

Determination of Titania

If the sample has a high chromic oxide content, giving a very yellow solution, the chromium in the control solution must be reduced by the addition of 2 ml of hydroxyammonium chloride solution (100 g/litre).

Transfer 20·0 ml of the stock solution to each of two 50-ml volumetric flasks, A and B. To each flask add 10 ml of diluted phosphoric acid (2 + 3) and, to flask A only, add 10 ml of hydrogen peroxide solution (6%). Dilute the solution in each flask to 50 ml and mix.

Measure the optical density of the solution in flask A against the solution in flask B in 40-mm cells at 398 nm, or by using a colour filter (Ilford 601) in a suitable instrument. The colour is stable for 5 min to 24 h after the addition of the hydrogen peroxide solution. Determine the titania content of the solution by reference to a calibration graph.

Determination of Manganese Oxide

Transfer 10·0 ml of the stock solution to a 250-ml beaker. Add 10 ml of diluted sulphuric acid (1 + 1), 5 ml of nitric acid ($d = 1·42$) and evaporate to strong fumes to remove chlorides. Add 20 ml of nitric acid ($d = 1·42$), 10 ml of diluted phosphoric acid (1 + 9) and about 50 ml of water. Boil to remove any nitrous fumes, cool slightly and add about 0·2 g of potassium periodate. Boil for 2 min, transfer to a steam bath and keep hot for 10 min. Allow to cool and transfer to a 100-ml volumetric flask. Dilute the solution in the flask to 100 ml and mix.

Measure the optical density of the solution against water in 40-mm cells at 524 nm, or by using a colour filter (Ilford 604) in a suitable instrument. Determine the manganese oxide content of the solution by reference to a calibration graph.

Determination of Chromic Oxide

Diphenylcarbazide method (for Cr_2O_3 contents up to approximately 0·1%)

Transfer 10·0 ml of the stock solution to a 100-ml beaker, add 5 ml of diluted sulphuric acid (1 + 9) and evaporate to dryness. To the dry residue add 2 ml of diluted sulphuric acid (1 + 9) and about 15 ml of water. Warm to dissolve as much of the residue as possible. Filter, if necessary, through a fine filter paper (Whatman No. 42 is suitable) and wash the residue with warm water. Evaporate to about 20 ml, if necessary. Add 2 ml of ammonium ceric nitrate solution (10 g/litre) and allow to stand on a steam bath for 25 min. Cool to 10°C and add sodium azide solution (20 g/litre), drop by drop, to destroy the colour of the excess ceric ion.

Transfer the solution to a 100-ml volumetric flask containing 3 ml of diluted sulphuric acid (1 + 9) and dilute to about 90 ml. Add 2 ml of diphenyl-carbazide solution (10 g/litre), dilute to 100 ml and mix. Allow to stand for 5 min.

Measure the optical density of the solution against water in 10-mm cells at 540 nm, or by using a colour filter (Ilford 605) in a suitable instrument. Determine the chromic oxide content of the solution by reference to a cali-bration graph.

EDTA method (for Cr_2O_3 contents above approximately 0·1%)

Transfer 50·0 ml of the stock solution to a separating funnel (a Squibb's type is recommended). Add 20 ml of chloroform and 10 ml of sodium diethyl-dithiocarbamate solution (100 g/litre). Stopper the funnel and shake vigor-ously. Release the pressure in the funnel by carefully removing the stopper and rinse the stopper and neck of the funnel with water. Allow the layers to separate and withdraw the chloroform layer. Add 10-ml portions of chloro-form and repeat the extraction until a coloured precipitate is no longer formed. Finally wash the aqueous phase at least three times with 10 ml portions of chloroform. Discard the chloroform extracts.

Transfer the aqueous phase to a 400-ml beaker, boil off traces of chloroform and cool to room temperature. Add 20 drops of hydrochloric acid ($d = 1\cdot18$), followed by 10 ml of sodium sulphite solution (50 g/litre) *with stirring*, and boil for 5 min. Cool to room temperature, add 10 ml of EDTA solution (50 g/litre), followed by ammonia solution ($d = 0\cdot88$), drop by drop, until the first appearance of a slight permanent precipitate. Dissolve this precipitate by adding 20 drops of diluted acetic acid ($1 + 1$) and dilute to about 200 ml. Heat the solution to boiling and boil for 10–15 min, cool, dilute to 250 ml in a volumetric flask and mix.

Measure the optical density of the solution against water in 40-mm cells at 550 nm, or by using a colour filter (Ilford 605) in a suitable instrument. Determine the chromic oxide content of the solution by reference to a calibration graph.

Determination of Alumina

Transfer 100·0 ml of the stock solution to a 400-ml beaker. Add 10 ml of sulphurous acid and boil off the excess sulphur dioxide. Cool slightly, add 5 ml of nitric acid ($d = 1\cdot42$) and boil for 15 min.

Cool to about 80°C, add 5 g of ammonium chloride and stir to dissolve; then add diluted ammonia solution ($1 + 1$), with stirring, until the solution is just alkaline to bromophenol blue. Boil off the slight excess of ammonia.

Allow the solution to stand for 5 min for the precipitate to settle, then filter through an open textured, hardened filter paper (Whatman No. 541 is suitable). Rinse the beaker with hot, faintly ammoniacal ammonium nitrate solution (10 g/litre) and pour the washings through the filter. Wash the precipitate well with hot, faintly ammoniacal ammonium nitrate solution (10 g/litre). Discard the filtrate and washings. Place the precipitation beaker under the funnel and dissolve the precipitate through the filter with 40 ml of hot, diluted hydrochloric acid ($1 + 1$). Wash the paper thoroughly with hot water and discard the paper.

Cool and transfer the solution to a separating funnel. The volume at this stage should be about 100 ml. Add 20 ml of chloroform and 10 ml of cupferron solution (60 g/litre). Stopper the funnel and shake vigorously. Release the pressure in the funnel by carefully removing the stopper and rinse the stopper and neck of the funnel with water. Allow the layers to separate and withdraw the chloroform layer. Confirm that extraction is complete by checking that the addition of a few drops of cupferron solution (60 g/litre) does not produce a permanent coloured precipitate. Add 10-ml portions of chloroform and repeat the extraction until the chloroform layer is colourless. Wash the stem of the funnel, inside and out with chloroform, using a polythene wash battle.

Discard the chloroform extracts and transfer the aqueous solution to a 500-ml conical flask. Add a few drops of bromophenol blue indicator and then add ammonia solution ($d = 0\cdot88$) drop by drop, until the solution is just alkaline. Re-acidify quickly with hydrochloric acid ($d = 1\cdot18$) and add 5–6 drops in excess. Add 10·0 ml of standard EDTA solution (0·05M); this is sufficient for 2·5% of alumina. Then add ammonium acetate buffer solution (for Al_2O_3 determination) until the indicator turns blue, followed by 10 ml in excess. Boil the solution for 10 min and cool. Add a volume of ethanol (95%) equal to the total volume of the solution and 1–2 ml of dithizone solution and titrate with the standard zinc solution (0·05M), using a semi-micro or similar burette, from blue/green to the first appearance of a permanent pink colour. [The pinkish tinge which sometimes appears in the solution after the addition of the dithizone and the purple colour due to the chromium –EDTA complex can both be screened out by the dropwise addition of Naphthol Green B solution (1 g/litre)].

If the EDTA solution is not exactly 0·05M, calculate the equivalent volume of exactly 0·05M EDTA solution.

If V is the volume of EDTA solution (0·05M) and
v ml is the volume of zinc solution used in back-titration, then
the percentage of alumina $= 0\cdot255\ (V-v)$.
Correct for the blank determination.

NOTE: The result must be corrected for any chromic oxide in the sample. Multiply the chromic oxide content expressed as a percentage by 0·667 and deduct this figure from the percentage of alumina.

Determination of Lime

Transfer a 50-ml aliquot of the stock solution to a 250-ml volumetric flask. Add 5 ml of diluted triethanolamine ($1 + 1$), 10 ml (for magnesites, or as appropriate for dolomites) of standard EGTA solution (0·05M approx.) and then dilute to 150 ml. Add potassium hydroxide solution (250 g/litre) until no further precipitation takes place and then add 10 ml in excess, folowed by 10 ml of Magflok solution (20 g/litre). Dilute to 250 ml, shake and allow to stand for about 10 min to settle. Filter through a 150-mm, *dry* coarse filter paper (Whatman No. 541 is suitable) into a *dry* beaker. Pipette 200 ml of the filtrate into a 500-ml conical flask and add 15 ml of potassium hydroxide solution (250 g/litre).

Add approximately 0·03 g of screened Calcein indicator and titrate with standard lime solution (0·05M) until the first appearance of green fluorescence.

The titration is best carried out in good daylight, but direct sunlight should be avoided.

If the EGTA solution is not exactly 0·05M, calculate the equivalent volume of exactly 0·05M EGTA solution.

If V ml of EGTA (0·05M) are taken and
v ml of CaO solution (0·05M) are used for back titration
then percentage CaO $= 0\cdot701 \times (\frac{4}{5}\ V-v)$.
Correct for the blank determination.

Determination of Magnesia

Transfer 100·0 ml of the "diluted stock solution" (see determination of ferric oxide) to a 500-ml separating funnel. Add diluted ammonia solution

$(1 + 1)$, drop by drop, until the solution is faintly alkaline to bromophenol blue. Just re-acidify with diluted hydrochloric acid $(1 + 3)$ and then add 4 ml in excess.

Add 20 ml of chloroform and 10 ml of sodium diethyldithiocarbamate solution (100 g/litre). Stopper the funnel and shake vigorously. Release the pressure in the funnel by carefully removing the stopper and rinse the stopper and neck of the funnel with water. Allow the layers to separate and withdraw the chloroform layer. (If an emulsion has formed, it will be necessary to add a few drops of hydrochloric acid and reshake).

Add 10-ml portions of chloroform and repeat the extraction with 5-ml portions of DDC until a *coloured* precipitate is no longer formed. Finally wash the aqueous phase three times with 10-ml portions of chloroform. This separation will remove iron and manganese.

If the sample contains more than approximately 1% alumina it will be necessary to carry out a buffered cupferron/chloroform solvent extraction as follows:

Add diluted ammonia solution $(1 + 1)$ dropwise until the solution is just alkaline to bromophenol blue. Just reacidify with diluted hydrochloric acid $(1 + 9)$ and add 20 ml of the ammonium acetate buffer solution (for MgO determination).

Add 20 ml of chloroform and 10 ml of cupferron solution (60 g/litre). Stopper the funnel and shake *vigorously*. Release the pressure in the funnel by carefully removing the stopper and rinse the stopper and neck of the funnel with water. Allow the layers to separate and withdraw the chloroform layer. Repeat the extraction with a further 10 ml of cupferron solution (60 g/litre) and finally wash the aqueous phase three times with 10-ml portions of chloroform. This separation will remove aluminium and titanium.

Transfer the aqueous phase from either the DDC separation, or, if alumina has been removed, from the buffered cupferron separation to a 500-ml conical flask and boil off traces of chloroform. Cool, and add 2 g of ammonium chloride and 2 ml of diluted triethanolamine $(1 + 1)$ with swirling followed by an appropriate known amount of standard DCTA solution (0·05M) (e.g. 80 ml for magnesites or 40 ml for raw dolomites). This addition is made to complex most of the magnesium before the solution is made alkaline, so that the tendency for magnesium hydroxide to precipitate is greatly reduced. Then add 30 ml of ammonia solution $(d = 0.88)$, followed by 5 ml of hydroxy-ammonium chloride solution (100 g/litre).

Add approximately 0·015 g of Solochrome Black 6B indicator and titrate with the standard DCTA solution (0·05M) from red through purple to the last change to a clear blue. This titration also includes the titration for lime in the sample, and this must be determined by the method given.

If the DCTA solution is not exactly 0·05M, calculate the equivalent volume of exactly 0·05M solution.

If V is the total amount of 0·05M DCTA added and

v is the equivalent amount of 0·05M DCTA required to react with the CaO in 0·2-g sample then percentage of MgO $= 1.008 \, (V - v)$.

Determination of Alkalis

Determine the alkalis in the sample by the procedure given in Chapter 16 under "Magnesites and Dolomites".

CHAPTER 12

ANALYSIS OF CHROME-BEARING MATERIALS

GENERAL

The chrome-bearing refractory is one of the more difficult types of ceramic materials to analyse. Decomposition is the first of many problems; few techniques yield consistent results: perchloric acid, potassium bisulphate, sodium peroxide and mixtures of sodium peroxide and hydroxide have all been recommended for this purpose, but all have proved somewhat variable in performance. The most consistently successful flux appears to be a mixture of alkali carbonate and boric acid, fairly rich in the latter.

The number of methods quoted previously have been reduced (in this volume) to two, one a modern method which has, for the most part, proved its capabilities in co-operative tests but which still needs to be handled intelligently, and the other the classical method, which, though very tedious, can yield good results in careful hands. It is still possible that the former may give rise to problems in circumstances which have not yet been encountered, but it has been used with success on a wide range of samples and will almost certainly form the basis of a future British Standard method.

The difficulties of decomposition mean that the minimum number of attacks are made, but the snags are such that four decompositions are still needed for the more recent method. The presence of four elements at major content level—iron, chromium, aluminium and magnesium, the very similar chemical nature of chromium and aluminium, and the ever present problems posed by large amounts of magnesium during precipitations give some indications of the complexities of any method. In addition, fairly high levels of accuracy are needed for minor amounts of silica and lime; in this matrix these two alone offer great difficulties.

Thus any analysis of a chrome-bearing material needs to be conducted carefully and with constant vigilance, even over and above that normally expected of the analyst. Even though the newer method is less prone to gross error, it is fair to say that as yet not all the possible snags have been met, noted and overcome.

PREPARATION OF SAMPLE

The sample prepared for analysis should be ground to pass completely through a 120-mesh B.S. test sieve. A non-metallic (e.g. nylon bolting-cloth) sieve is preferable.

The preparation of a ground sample of a chrome-bearing material is complicated by the fact that the material is often itself magnetic and therefore precludes the use of a magnet to remove iron contamination. However, unless materials such as boron carbide are available, there is little alternative to completing the crushing in an iron percussion mortar. Fortunately the iron content of these materials is so high that the iron introduced represents only a small fraction of the amount in the sample.

The material is too hard in general to permit the use of either porcelain or agate mortars.

A. Classical Method

PRINCIPLE OF THE METHOD

The method, in principle, follows the normal pattern of classical analysis, with the necessary modifications to take account of the presence of chromium and large amounts of magnesia in the sample. Only 0·5 g of sample is taken, owing to the difficulty of handling large ammonia group precipitates, but it should be noted that this results in an immediate doubling of weighing errors.

The sample is fused in sodium carbonate and boric acid over a gas burner, at first with a low heat and slowly raising the temperature to the maximum consistent with maintaining oxidizing conditions. If reduction starts, as indicated by the formation of small, black, shiny metallic crystals in the melt, accompanied by a reduction of the yellow colour, it is almost impossible to reverse. The fusion should be discontinued immediately otherwise considerable damage may be caused to the platinum ware.

The melt is dissolved in diluted hydrochloric acid, care being taken to ensure that the acid is not sufficiently strong to cause reduction of the dichromate and production of chlorine which will attack the platinum. The silica is determined by double evaporation to dryness and filtration, followed by ignition and weighing before and after treatment with hydrofluoric and sulphuric acids. Platinum dishes cannot be used for these evaporations. The residue from the hydrofluoric acid treatment is fused in sodium carbonate and combined with the main filtrate.

Although the bulk of the chromium is reduced by the successive evaporations with hydrochloric acid, some Cr^{VI} may still be present. This would escape the ammonia precipitation and is therefore reduced with sulphurous acid. During this process the iron is reduced to the ferrous state and needs re-oxidizing, nitric acid being selected in preference to bromine as the latter would also oxidize the manganese which would then be precipitated. It is preferable to precipitate the manganese in the "catch" precipitation because its amount, and thus completeness of precipitation, can be more readily judged.

After the reduction and re-oxidation stages, the ammonia group oxides (Fe, Ti, Mn, Cr, Al) are precipitated. The nature and bulk of the precipitate, the presence of large amounts of magnesia and, of course, alkali salts and boric acid, all tend to cause co-precipitation. For this it is necessary to carry out three precipitations in all and, after evaporating the bulk of the liquid, to carry out a "catch" precipitation with the manganese oxidized with bromine. Even this precipitation has to be repeated to ensure freedom from co-precipitation. Considerable care needs to be taken throughout these steps otherwise the complete recovery of an uncontaminated precipitate is impossible. The precipitate is ignited and weighed; the ignition is again a compromise. At

1000°C there is little doubt that the results tend to be high as the alumina, at least, will lose further weight if ignited to 1200°C, whereas at 1200°C there is a danger that the ferric oxide, present in considerable amount, may start losing oxygen. It is thought that the lesser risk is entailed by igniting to the higher temperature.

The ignited precipitate is fused in sodium carbonate and the melt extracted with water. Aluminium and chromium pass into solution as aluminate and chromate and are filtered off. To prevent their being accompanied by manganese this is reduced by a few millilitres of alcohol. Any silica remaining in solution after double dehydration will accompany the chromium and aluminium. Two fusions are necessary to ensure complete separation. The residue is fused in potassium pyrosulphate and the melt dissolved in dilute sulphuric acid. Titania is determined on this solution with hydrogen peroxide, and manganese with periodate, both spectrophotometrically. Alternative methods are described for iron, either spectrophotometrically with 1:10 phenanthroline after reduction with hydroxyammonium chloride, or by titration with dichromate after reduction with stannous chloride. The former method is the more simple and rapid, and experience has shown that by careful use of a good spectrophotometer, results are equally as good as by the volumetric method.

The alkaline solution containing chromate is used, in the method, solely for the determination of Cr_2O_3. The solution is boiled alkaline with sodium peroxide (for long enough to destroy the excess peroxide) and then acid with permanganate to ensure that all the chromium is present as Cr^{vi}. A small amount of hydrochloric acid is used to destroy the excess permanganate. The solution is cooled, the chromate reduced with ferrous ammonium sulphate and the excess ferrous ions titrated with dichromate solution.

The alumina content is then obtained by calculation.

It is probably better to determine chromium on a separate sample as described in the second method in this chapter and to use the sodium carbonate extract for the determination of alumina and possibly residual silica by methods similar to those described in the second method. Care would need to be taken to use vessels for the extraction of the melts from which silica would not be extracted (e.g. platinum or polythene).

The filtrate from the ammonia group oxide precipitation is evaporated to dryness with nitric acid to remove ammonium salts which tend to interfere in the determination of lime and magnesia. Ideally, this process should be repeated after the determination of lime, but the large content of magnesia tends to be little affected by ammonium salts especially as their concentration has been greatly reduced by this first evaporation. The residue is dissolved and the lime precipitated as oxalate. A double precipitation is necessary and if magnesia has been precipitated in any quantity during the first precipitation (as seen from precipitate adhering to the sides of the beaker), a third precipitation is highly desirable to ensure the complete separation of lime and magnesia. As the weight of lime is low, the accuracy of the determination is probably better if the oxalate is titrated with permanganate after dissolving in sulphuric acid than if the precipitate is ignited and weighed.

Magnesium is determined as the pyrophosphate by igniting the precipitate from alkaline phosphate precipitation. Two precipitations and care over the details of the procedure are necessary if the correct weighing form is to be obtained. Ignition should be carried out at 1150°C, as at this temperature precipitates can almost be guaranteed to burn to a pure white.

Alkalis are determined by flame photometer after decomposition with hydrofluoric, nitric and sulphuric acids. The samples are difficult to decompose, and much of the sample remains apparently unattacked, so that the absolute accuracy of the alkali results must remain in some doubt. The determination, nevertheless, has a value in ensuring the absence of abnormal amounts of alkali, or detecting the presence of an alkali bond. In the latter case as the alkali is not truly bound in the sample, the results are very probably sound.

REAGENTS

Additional Reagents

Barium diphenylamine sulphonate
Mercuric chloride
Sulphur dioxide: liquid
Tin: granulated

Prepared Reagents

Ammonium acetate (approx. 10%): Dilute 140 ml of acetic acid (glacial) to 2000 ml with water and add carefully 140 ml of ammonia solution ($d = 0·88$), mix, cool and adjust to approximately pH 6, either with acetic acid or ammonia solution.

Ammonium nitrate (10 g/litre): Dilute 10 ml of nitric acid ($d = 1·42$) to about 200 ml. Add diluted ammonia solution ($1 + 1$) until the solution is faintly alkaline to methyl red. Dilute to 1 litre.

Ammonium phosphate (100 g/litre): Dissolve 2 g of di-ammonium hydrogen orthophosphate in 20 ml of water. This solution should be freshly prepared.

Bromine water (saturated): Shake 20 ml of bromine with 500 ml of water in a glass stoppered bottle.

Mercuric chloride (saturated): Shake 20 g of mercuric chloride with 250 ml of water. Allow to stand 24 h before use.

1:10 phenanthroline (10 g/litre): Prepare enough solution for immediate use at a concentration of 0·1 g of 1:10 phenanthroline hydrate in 10 ml of diluted acetic acid ($1 + 1$).

Silver nitrate (1 g/litre): Used for testing for chlorides: Dissolve 0·1 g of silver nitrate in 100 ml of diluted nitric acid ($1 + 9$).

Stannous chloride (30 g/litre): Dissolve by warming, 3 g of a stannous chloride in 30 ml of hydrochloric acid ($d = 1·18$). Cool and dilute to 100 ml. Store in a rubber-stoppered bottle, with a small piece of tin in the solution to prevent oxidation.

Sulphuric–phosphoric acid mixture: Cautiously add 75 ml of sulphuric acid ($d = 1\cdot84$) and 75 ml of orthophosphoric acid ($d = 1\cdot75$) to 350 ml of water, cooling the solution and keeping it well mixed during the addition of the acids.

Sulphurous acid: Saturate 250 ml of water with sulphur dioxide.

Indicators

Barium diphenylamine sulphonate
Bromophenol blue
Methyl red

Standard Solutions

Ferrous ammonium sulphate (approx. $0\cdot1$N)
Potassium dichromate ($0\cdot1$N)
Potassium dichromate ($0\cdot05$N)
Potassium permanganate (approx. $0\cdot1$N)

PROCEDURE

Determination of Loss on Ignition

Weigh $1\cdot000$ g of the finely ground, dried (110°C) sample into a platinum crucible. Place the crucible in a muffle furnace and slowly raise the temperature to 1000 ± 25°C. Ignite to constant weight at this temperature; 30 min ignition is usually sufficient.

For routine work it is possible to start the ignition over a low mushroom flame, slowly increasing the temperature to full heat over a period of about 20 min after which the crucible is transferred to a furnace at 1000°C for 30 min.

Determination of Silica, Alumina, Ferric Oxide, Titania, Manganese Oxide, Chromium Sesquioxide, Lime and Magnesia

Decomposition of the Sample

Weigh $0\cdot500$ g of the finely ground, dried (110°C) sample into a platinum crucible. Add 4 g of anhydrous sodium carbonate and $1\cdot5$ g of boric acid and mix thoroughly.

Heat over a Bunsen or Méker flame, slowly raising the temperature until the mixture begins to melt, and keep at this temperature until melting is complete. It is important that the initial stage of heating should be slow, as a vigorous reaction takes place on melting and spattering easily occurs. Again raise the temperature slowly and steadily to the full heat of the flame (approx. 950°C).

After about 5 min at this temperature, swirl the contents of the crucible every 2 min, making sure that the particles of sample on the sides of the crucible come into contact with hot, molten flux.

If swirling is begun at too early a stage, it is difficult to detach unfused particles from the sides of the crucible.

Continue heating at this temperature for 25–45 min depending on the nature of the sample. Completeness of decomposition can be checked visually by the absence of any black, unfused material at the bottom of the melt.

The type and position of the burner should be chosen so as to maintain oxidizing conditions throughout the process.

Cool, place the crucible and lid in a 250-ml beaker containing 75 ml of water and cover the beaker with a clock glass. Introduce 25 ml of hydrochloric acid ($d = 1\cdot18$) through the lip of the beaker.

Stand on a steam bath until disintegration of the melt is complete. Then remove the crucible and lid, washing them with a jet of hot water and scrubbing them with a "bobby".

If any appreciable amount of residue is left, owing to incomplete fusion, this may be filtered off, ignited and re-fused.

Determination of Silica

Transfer the solution to a porcelain evaporating basin and add 25 ml of methanol. Place the basin on a steam bath and gently evaporate until most of the alcohol has been removed. Evaporate to dryness.

Cool, drench the residue with 10 ml of hydrochloric acid ($d = 1\cdot18$) and allow to stand for a few minutes. Add 75 ml of hot water and digest on a steam bath for 5 min to dissolve the salts.

Filter through a No. 40 Whatman paper, transferring the precipitate to the filter with a jet of hot, diluted hydrochloric acid (1 + 9) and scrubbing the basin with a "bobby". Wash five times with hot, diluted hydrochloric acid (1 + 19) and then with hot water until the washings are free from chlorides. Reserve the paper and residue for the subsequent ignition and transfer the filtrate and washings back to the evaporating basin.

Add a further 25 ml of methanol and again evaporate completely to dryness on a steam bath. Cover the basin with a clock glass and bake in an air oven at 110°C for 1 h.

Allow to cool, then drench the residue with 10 ml of hydrochloric acid ($d = 1\cdot18$). Allow to stand for a few minutes, add about 75 ml of hot water and digest on a steam bath for 5 min to dissolve the salts.

Filter through a No. 42 Whatman paper, transferring the residue to the filter with a jet of hot, diluted hydrochloric acid (1 + 19) and scrubbing the basin with a "bobby". Wash five times with hot, diluted hydrochloric acid (1 + 19) and then with hot water until it is free from chlorides.

Reserve the filtrate and washings for the determination of the ammonia group oxides.

Place the two residues and papers, without drying, in a platinum crucible and heat cautiously to dry the residues and char the papers. Then burn off the carbon at a low temperature, otherwise it may be impossible to remove all of it, and finally ignite at 1200°C for 15 min and then to constant weight. Allow to cool and weigh to obtain the weight of the impure silica.

Moisten the weighed residue with 10 drops of diluted sulphuric acid (1 + 1) and add about 5 ml of hydrofluoric acid (40% w/w). Evaporate to dryness on a sand bath in a fume cupboard. Ignite the dry residue at 1200°C for 5 min, allow the crucible to cool and weigh.

Subtract the weight of the residue from the weight of the impure silica to obtain the weight of silica in the sample taken. If the residue weighs more than 5 mg repeat the treatment with sulphuric and hydrofluoric acids to ensure that all the silica has been removed.

Fuse the residue with about 1 g of anhydrous sodium carbonate and dissolve the melt in diluted hydrochloric acid (1 + 1). Add the solution to that reserved for the determination of the ammonia group oxides.

Determination of Total Ammonia Group Oxides (R_2O_3)

Adjust the volume of the solution reserved for the determination of ammonia group oxides to about 300 ml, add 10 ml of sulphurous acid and boil off the excess sulphur dioxide.

Cool the solution slightly, add 5 ml of nitric acid ($d = 1.42$) and boil for 10–15 min. Cool to about 80°C, add 3–5 g of ammonium chloride followed by diluted ammonia solution (1 + 1), with stirring, until precipitation appears to be complete.

Make the solution just alkaline and boil for 2 min, allow to stand for 5 min for the precipitate to settle, and filter through a No. 41 Whatman paper.

Transfer the precipitate to the filter and wash five times with hot, faintly ammoniacal ammonium nitrate solution (10 g/litre). Just acidify the filtrate and washings with hydrochloric acid ($d = 1.18$) and reserve.

Place the precipitation beaker under the funnel and dissolve the precipitate through the paper with hot, diluted hydrochloric acid (1 + 3) and wash the paper thoroughly with hot water. (The same paper is used for filtering the precipitate from the second precipitation.)

Dilute the solution to about 300 ml and repeat the precipitation as before. Filter through the previously used paper, transfer the precipitate to the filter and wash three times with hot, faintly ammoniacal ammonium nitrate solution (10 g/litre).

Just acidify the filtrate and washings with hydrochloric acid ($d = 1.18$) and reserve; transfer the precipitate and paper back to the precipitation beaker.

Dissolve the precipitate in a slight excess of hydrochloric acid ($d = 1.18$) and macerate the filter paper. Dilute the solution to about 250 ml and repeat the precipitation as before.

Filter through a No. 541 Whatman paper, transferring the precipitate to the filter with hot, faintly ammoniacal ammonium nitrate solution (10 g/litre) and scrubbing the beaker with a "bobby". Wash the precipitate free from chlorides with hot, faintly ammoniacal ammonium nitrate solution (10 g/litre).

Just acidify the filtrate and washings with hydrochloric acid ($d = 1.18$) and add to the reserved solutions; retain the paper and precipitate for ignition.

Evaporate the combined filtrates to about 250 ml and add about 5 ml of bromine water. Make just alkaline to bromophenol blue with diluted ammonia solution (1 + 1) and boil off the excess ammonia.

Filter through a No. 40 Whatman paper, transferring the precipitate to the filter with a jet of hot, faintly ammoniacal ammonium nitrate solution (10 g/litre). Wash the precipitate five times with hot, faintly ammoniacal ammonium nitrate solution (10 g/litre).

Just acidify the filtrate and washings with hydrochloric acid ($d = 1\cdot18$) and reserve for the determination of lime; transfer the precipitate and paper back to the precipitation beaker. Dissolve the precipitate in a slight excess of hydrochloric acid ($d = 1\cdot18$) and macerate the filter paper.

Dilute to about 150 ml, add about 5 ml of bromine water and repeat the precipitation as before.

Filter through a No. 540 Whatman paper and wash the precipitate free from chlorides with hot, faintly ammoniacal ammonium nitrate solution (10 g/litre).

Reserve the precipitate and paper for ignition; just acidify the filtrate and washings with hydrochloric acid ($d = 1\cdot18$) and add to the solution reserved for the determination of lime.

Place the precipitates and papers, reserved for ignition, in a weighed platinum crucible and heat, slowly at first, to dry the precipitates and char the papers. Burn off the carbon and finally ignite at 1200°C to constant weight; 30 min is usually sufficient. Cool and weigh.

The weight of the mixed ammonia group oxides is thus obtained.

Separation of the Mixed Ammonia Group Oxides

Transfer the ignited oxides to an agate mortar and mix thoroughly with 5 g of anhydrous sodium carbonate and 0·1 g of potassium nitrate. Return the mixture to the crucible and "rinse" out the mortar with a further 2 g of anhydrous sodium carbonate. Add this to the contents of the crucible and mix thoroughly.

Fuse the mixture over a Méker burner; raise the temperature very slowly until frothing ceases, then complete the fusion at about 900°C for 1 h, occasionally swirling the melt to ensure thorough mixing.

Quench the melt by immersing the bottom half of the crucible in cold water, then place the crucible and lid in about 100 ml of hot water contained in a 250-ml beaker. Add 1–2 ml of ethanol (95%) to reduce the manganese, cover the beaker with a clock glass and stand it on a steam bath until the melt is completely disintegrated.

Remove the crucible and lid, washing them thoroughly with hot water, and scrubbing them with a "bobby". Crush any lumps remaining in the solution and filter through a No. 42 Whatman paper. Transfer the residue to the filter with a jet of hot sodium carbonate solution (10 g/litre), scrubbing the beaker with a "bobby". Wash eight times with hot sodium carbonate solution (10 g/litre) and reserve the filtrate and washings for the determination of chromium sesquioxide.

Transfer the residue and paper to the crucible and heat slowly to dry the residue and char the paper; then burn off the carbon at a low temperature.

Mix the residue in the crucible with about 5 g of anhydrous sodium carbonate and fuse for about 30 min at 900°C.

Extract with water, add ethanol, filter and wash as before. Add the filtrate and washings to those already reserved for the determination of chromium sesquioxide.

Again return the residue and paper to the crucible and heat slowly to dry the residue and char the paper; then burn off the carbon at as low a temperature as possible.

Carefully fuse the residual oxides in the same platinum crucible with about 10 g of potassium bisulphate. Cool the crucible and extract the melt in a 250-ml beaker containing about 150 ml of water to which 10 ml of sulphuric acid ($d = 1 \cdot 84$) has been carefully added.

Cool the solution and dilute to 250 ml in a volumetric flask. The solution thus obtained will be referred to as the "stock test solution".

Determination of Ferric Oxide

Two methods are available for the determination of ferric oxide, spectrophotometric or volumetric. The results of each are comparable and appear to be of equal accuracy.

Spectrophotometric method.

Transfer a 25-ml aliquot of the stock test solution to a 200-ml volumetric flask, dilute to 200 ml and mix.

Transfer a 20-ml aliquot of this diluted solution to a 200-ml volumetric flask. Add 2 ml of hydroxyammonium chloride solution (100 g/litre), 5 ml of 1:10-phenanthroline solution (10 g/litre) and ammonium acetate solution (approx. 10%) until a pink colour appears in the solution followed by 2 ml in excess.

Allow to stand for 15 min, dilute to 200 ml and mix. Measure the optical density of the solution against water in 10-mm cells at 510 nm in a suitable instrument. (The ferric oxide content is too high in most of these materials to permit the use of a filter-type instrument.)

Stand the solution in the cells for a further 5 min and check that there has been no increase in optical density. Repeat, if necessary, to constant optical density.

Determine the ferric oxide content of the solution by reference to a calibration graph.

Using the Unicam SP600 spectrophotometer, the calibration graph extends to about 20% Fe_2O_3. If this amount is exceeded the dilution of the solution can be adjusted to give a suitable optical density and the ferric oxide content calculated accordingly.

Volumetric method

Take as large an aliquot as possible (usually 125-ml of the stock test solution. Heat to 80°C, add diluted ammonia solution (1 + 1) until the solution is slightly alkaline to methyl red and filter through a No. 541 Whatman paper. Wash several times with hot water, then wash the bulk of the precipitate back into the precipitation beaker.

Dissolve with hydrochloric acid ($d = 1 \cdot 18$) any precipitate remaining on the paper and wash thoroughly with hot water into the beaker containing the main precipitate. Make up the total volume of hydrochloric acid used to 25 ml and warm to dissolve the remainder of the precipitate.

Transfer the solution to a conical flask and boil. While boiling, add

F*

stannous chloride solution (30 g/litre), drop by drop, until the solution becomes colourless; then add 1–2 drops in excess.

Cool rapidly to room temperature, keeping the flask covered with a clock glass to minimize the risk of oxidation and add 5 ml of mercuric chloride solution.

Allow to stand for 10 min, add 15 ml of sulphuric acid–phosphoric acid mixture and 10 drops of barium diphenylamine sulphonate indicator. Titrate to a purple colour with standard potassium dichromate solution (0·05N).

Calculation—1 ml 0·05N $K_2Cr_2O_7 \equiv 0·00399$ g Fe_2O_3.

Determination of Titania

Transfer a 40-ml aliquot of the stock test solution to each of two 100-ml volumetric flasks, A and B. To each flask add 10 ml of diluted phosphoric acid (2 + 3) and, to flask A only, 10 ml of hydrogen peroxide solution (6%). Dilute the solution in each flask to 100 ml and shake well.

Measure A against B in 40-mm cells at 398 nm or by using a colour filter (Ilford 601) in a suitable instrument. The colour is stable from 5 min until 24 h after the addition of the hydrogen peroxide solution.

Determine the titania content of the solution by reference to a calibration graph.

Determination of Manganese Oxide

Transfer a 50-ml aliquot of the stock test solution to a 250-ml beaker. Add 20 ml of nitric acid ($d = 1·42$) and 10 ml of diluted phosphoric acid (1 + 9).

Boil to remove nitrous fumes, cool slightly and add about 0·2 g of potassium periodate.

Boil for 2 min, transfer to a steam bath and keep hot for 10 min.

Allow to cool and transfer to a 100-ml volumetric flask. Dilute the solution to 100 ml and mix.

Measure the optical density of the solution against water in 40-mm cells at 524 nm or by using a colour filter (Ilford 604) in a suitable instrument.

Determine the manganese oxide content of the solution by reference to a calibration graph.

Determination of Chromium Sesquioxide

Boil down the combined alkaline filtrates reserved for the determination of chromium sesquioxide to about 100 ml. Cool and add approximately 0·5 g of sodium peroxide. Boil the solution for 10 min to decompose the excess of peroxide.

Cool and cautiously add 120 ml of diluted sulphuric acid (1 + 3). Transfer the solution to a 1-litre conical flask and dilute to about 500 ml.

Add potassium permanganate solution (approx 0·1N), drop by drop, sufficient to give an excess as shown by the persistence of an amber colour in the solution after boiling for 5 min. Cool slightly, add 20 ml of diluted hydrochloric acid (1 + 3) and continue boiling for 15–20 min to destroy the excess permanganate and remove chlorine.

Cool rapidly to room temperature and reduce the dichromate by adding a weighed excess of ferrous ammonium sulphate crystals.

The practice of adding solid ferrous ammonium sulphate appears to be quite satisfactory for routine control purposes, provided that certain obvious precautions are taken. The crystals should be of uniform grade and should be thoroughly mixed before use. About 3 g of the salt should be checked periodically against standard potassium dichromate solution (0·1N). On the other hand if a 0·1N solution of ferrous ammonium sulphate is used this must be checked each time.

Add 25 ml of phosphoric acid ($d = 1·75$) and 10 drops of barium diphenylamine sulphonate indicator, and back-titrate the excess ferrous ammonium sulphate with standard potassium dichromate solution (0·1N) until the green colour changes to blue-green and then to purple.

The volume of standard potassium dichromate solution required to oxidize the same weight of ferrous ammonium sulphate as used in the determination is ascertained via a similar titration.

Calculation—Let V ml be the volume of standard potassium dichromate solution (0·1N) required to oxidize 1 g of ferrous ammonium sulphate, and w g be the weight of ferrous ammonium sulphate added to the chromate solution. Then if v ml of standard potassium dichromate solution (0·1N) were required to back-titrate the excess of ferrous ammonium sulphate, and 0·5 g of material was taken originally, the percentage of Cr_2O_3, x, in the sample is given by:

$$x = (wV - v) \times 0·5067$$

Determination of Alumina

The weight of alumina in the sample is obtained by calculation.

Subtract the weights of chromium sesquioxide, ferric oxide, titania and manganese oxide from the weight of the mixed oxides obtained from the ammonia precipitation to obtain the weight of alumina in the sample.

Determination of Lime

Evaporate the filtrate reserved for the determination of lime to about 300 ml. Cool, cover with a clock glass, cautiously add about 100 ml of nitric acid ($d = 1·42$) through the lip of the beaker and evaporate to dryness, keeping the beaker covered throughout the evaporation. (It is possible that a further treatment with nitric acid will be required depending on the amount of ammonium salts in the solution.)

Cool, add 5 ml of hydrochloric acid ($d = 1·18$), dilute to about 300 ml and warm until the salts have dissolved.

Add 5 g of oxalic acid and boil the solution, then add, with stirring, diluted ammonia solution (1 + 1) until the solution is alkaline to bromophenol blue and then add an excess of 10 ml of diluted ammonia solution (1 + 1). Cover the beaker with a clock glass and digest on a steam bath for 2 h.

Cool rapidly and filter through a Buchner funnel with a sintered-glass mat (porosity 4) or a No. 42 Whatman paper and wash the precipitate eight times with cold ammonium oxalate solution (10 g/litre).

Just acidify the filtrate and washings and reserve for the determination of magnesia. Dissolve the precipitate through the filter with 20 ml of hot, diluted nitric acid (1 + 1) and wash thoroughly with hot water.

Transfer the solution and washings back to the precipitation beaker, add about 2 g of oxalic acid, boil the solution and precipitate the calcium as before.

Cool rapidly, filter through a Buchner funnel with a sintered-glass mat (porosity 4) or a No. 42 Whatman paper and wash the precipitate thoroughly with cold water.

Just acidify the filtrate and washings with hydrochloric acid ($d = 1.18$) and reserve for the determination of magnesia.

If there is visible evidence of the precipitation of magnesium oxalate (i.e. precipitate adhering to the glass) in the first precipitation the second calcium precipitate must be redissolved, reprecipitated, filtered and washed as described previously, the filtrate and washings being reserved for the determination of magnesia.

Dissolve the second or third calcium precipitate, as the case may be, through the filter with 50 ml of hot, diluted sulphuric acid (1 + 9) and wash the filter thoroughly with hot water.

Transfer the solution and washings back to the precipitation beaker, heat to boiling and titrate with standard potassium permanganate solution (approx. 0.1N).

Calculation—1 ml 0.1N $KMnO_4$ $\equiv$ 0.002804 g CaO

Determination of Magnesia

Combine the filtrates from the various precipitations of lime and evaporate to about 350–400 ml.

Add 5 g of di-ammonium hydrogen phosphate, the solution must remain acid until the actual precipitation is carried out, hydrochloric acid ($d = 1.18$) being added if necessary.

Heat the solution to boiling and add diluted ammonia solution (1 + 1), with stirring, until the solution is just alkaline to bromophenol blue and then cool to room temperature.

Add 30 ml of ammonia solution ($d = 0.88$), stir vigorously to start the precipitation and allow to stand overnight at a low temperature.

Filter off the magnesium phosphate precipitate on a No. 42 Whatman paper.

Transfer the precipitate to the filter with a jet of cold, diluted ammonia solution (1 + 39). Wash the precipitate in the filter four times with cold, diluted ammonia solution (1 + 39), discarding the filtrate and washings.

Dissolve the precipitate through the filter with 30 ml of hot, diluted hydrochloric acid (1 + 1) and wash thoroughly with hot water. Transfer the solution and washings back to the beaker which held the main precipitate.

Add 10 ml of ammonium phosphate solution (100 g/litre), heat the solution to boiling and, while stirring, make just alkaline to bromophenol blue with diluted ammonia solution (1 + 1).

Cool, add 20 ml of ammonia solution ($d = 0.88$) and stir vigorously to start the precipitation. Stand overnight at a low temperature.

Filter through a No. 42 Whatman paper, transferring the precipitate to the paper with a jet of cold, diluted ammonia solution (1 + 39) and scrubbing the beaker with a "bobby". Wash thoroughly with cold, diluted ammonia solution (1 + 39), discarding the filtrate and washings.

Transfer the precipitate and paper to a weighed platinum crucible and heat cautiously to dry the precipitate and char the paper. Burn off the carbon at a low temperature. (Platinum crucibles may be seriously attacked during the ignition if the precipitate is not thoroughly washed with diluted ammonia solution, or if the precipitate is ignited too strongly before all the carbon is oxidized.) Finally ignite at 1150°C for 15 min and then to constant weight. Weigh the residue as magnesium pyrophosphate.

Calculation—Weight of $Mg_2P_2O_7 \times 0.3623$ = weight of MgO.

Determination of Alkalis

Determine the alkalis in the sample by the procedure given in Chapter 16 under "Chrome-bearing Materials".

B. Proposed British Standard Method

The following method is, at the time of writing, in the late stages of testing for submission to the British Standards Institution. Most of the determinations have already received the full approval of the committee.

PRINCIPLE OF THE METHOD

Loss on Ignition

The loss on ignition is determined at 1000°C.

Determination of Silica

The sample is fused in fusion mixture/boric acid and dissolved in sulphuric acid. After dilution, ammonium molybdate is added and the yellow silicomolybdic acid allowed to develop. After adjustment of acidity the silicomolybdic acid is precipitated by the addition of quinoline, the solution heated to 80°C and then cooled to below 20°C. The precipitate is then filtered on a weighed Gooch crucible, washed and dried at 150°C in an air-oven.

Determination of Chromium Sesquioxide

The sample is fused in fusion mixture/boric acid and dissolved in sulphuric acid. The solution is transferred to a 1-litre conical flask and boiled. Dilute potassium permanganate solution is added, to oxidize any reduced chromate, until the reddish-amber colour persists, and the excess permanganate is then destroyed by boiling with dilute hydrochloric acid. The solution is then cooled, excess standard ferrous ammonium sulphate added, followed by phosphoric acid. The excess ferrous ammonium sulphate is back-titrated with standard potassium dichromate using barium diphenylamine sulphonate as indicator.

Determination of Titania, Ferric Oxide, Alumina, Manganese Oxide, Lime and Magnesia

The sample is fused in fusion mixture/boric acid and dissolved in sulphuric acid. Chromium is then removed by extraction with Amberlite LA-2 liquid ion-exchange resin. The aqueous phase is washed by extraction with chloroform, transferred to a beaker and all traces of chloroform removed by boiling. After cooling, the solution is made up to standard volume. Aliquots of the "stock solution" are used for all following determinations.

The titania is determined spectrophotometrically with hydrogen peroxide, the ferric oxide with 1:10-phenanthroline and the manganese oxide with potassium periodate. For the alumina determination an aliquot is extracted with sodium diethyldithiocarbamate/chloroform to remove iron and manganese, followed by a cupferron/chloroform extraction to remove titanium. The determination is completed volumetrically with diaminocyclohexanetetraacetic acid (DCTA) and zinc, using dithizone as indicator. Any residual chromium present in the stock solution does not titrate, as the solution is not boiled after addition of DCTA. Lime is determined on a measured volume of the solution to which triethanolamine is added to complex interfering elements. A known volume of standard EGTA is then added: this is sufficient to complex all the calcium, and the magnesium is then precipitated with potassium hydroxide. After filtration the excess EGTA is titrated with standard calcium solution, Calcein being used as indicator. Magnesia is determined volumetrically on an aliquot of the stock solution by extraction with sodium diethyldithiocarbamate/chloroform to remove iron and manganese, followed by a buffered cupferron/chloroform solvent extraction to remove aluminium and titanium. After addition of triethanolamine, the magnesia is titrated in a strongly ammonical solution containing ammonium chloride, with DCTA. This titration also includes the titration for lime which must be allowed for.

Determination of Alkalis

The sample is decomposed with hydrofluoric, nitric and sulphuric acids to remove the silica. A further evaporation with nitric and sulphuric acids removes traces of flouride; the residue is dissolved in nitric acid and filtered. After dilution to a standard volume the alkalis are determined using a flame photometer.

REAGENTS

Additional Reagents

Amberlite LA-2 resin
Barium diphenylamine sulphonate
Magflok
Quinoline (boiling range 235–239°C)

Prepared Reagents

Amberlite LA-2 resin (1 + 4): Add 50 ml of resin to 200 ml of chloroform and mix. Transfer to a 500-ml separating funnel, add 25 ml of diluted sulphuric acid (1 + 9) and shake gently **(release the pressure in the funnel frequently during the first few seconds of shaking)**. Allow the layers to separate and run off the organic layer into a 250-ml measuring cylinder.

Ammonium acetate (approx. 10%): Dilute 140 ml of acetic acid (glacial) to 2000 ml with water and add carefully 140 ml of ammonia solution ($d = 0.88$). Mix, cool and adjust to approximately pH 6.0–6.5 either with acetic acid or ammonia solution.

Ammonium acetate buffer (for Al_2O_3 determination): Add 120 ml of acetic acid (glacial) to 500 ml of water followed by 74 ml of ammonia solution ($d = 0.88$). Mix, cool and dilute to 1 litre.

Ammonium acetate buffer solution (for MgO determination): Dilute 5 ml of ammonia solution ($d = 0.88$) to 100 ml with distilled water and add 30 ml of acetic acid (glacial). Adjust to pH 3.8 (if necessary) using a pH meter and then dilute to 200 ml with distilled water.

Ammonium molybdate (100 g/litre): Dissolve 100 g of ammonium molybdate in 1 litre of water; filter if necessary. Store in a polythene bottle and discard after 4 weeks or earlier if any appreciable deposit of molybdic acid is observed.

Cupferron (60 g/litre): Dissolve 3 g of cupferron in 50 ml of cold water and filter. This solution must be freshly prepared.

If the reagent is discoloured or gives a strongly coloured solution a new stock should be obtained. The solid reagent should be stored in a tightly stoppered bottle in the presence of a piece of ammonium carbonate to prevent decomposition.

Magflok (20 g/litre approx.): The resin is a thick viscous liquid and it is therefore most convenient to transfer a drop to a beaker and to weigh the amount taken. Sufficient water is then added to make up a solution approx. 20 g/litre.

This material is a partially hydrolysed polyacrylamide and is available from Ridsdale & Co. Limited, Newham Hall, Middlesbrough, under the trade name Magflok.

1:10-Phenanthroline (10 g/litre): Prepare enough solution for immediate use at a concentration of 0.1 g of 1:10-phenanthroline hydrate in 10 ml of diluted acetic acid (1 + 1).

Quinoline (2% v/v): Add 20 ml of quinoline (boiling range 235–239°C) to about 800 ml of hot water acidified with 25 ml of hydrochloric acid ($d = 1.18$) stirring constantly if cloudy. Cool, add some paper pulp and stir vigorously. Allow to settle and filter under suction through a paper pulp pad to remove traces of oily matter, but do not wash. Dilute the filtrate to 1 litre.

Quinoline wash solution (0.05% v/v): Dilute 25 ml of the quinoline solution (2% v/v) to 1 litre.

Sodium diethyldithiocarbamate (100 g/litre): Dissolve 5 g of sodium diethyldithiocarbamate in 50 ml of water and filter. This solution must be freshly prepared.

Indicators

Barium diphenylamine sulphonate
Bromophenol blue
Calcein (screened)
Dithizone
Solochrome Black 6B

Standard Solutions

Calcium (1·0 mg CaO/ml)
DCTA (0·05M)
EGTA (0·05M)
Ferrous ammonium sulphate (approx. 0·1N)
Potassium dichromate (0·05N)
Zinc (0·05M)

BLANK DETERMINATIONS

A blank determination shall be carried out on all reagents in accordance with the general scheme of analysis. When carrying out the blank determination for lime, it is necessary to add 50 ml of the magnesium sulphate heptahydrate solution (6 g/litre) before the addition of the standard EDTA solution.

PROCEDURE

Determination of Loss on Ignition

Weigh 1·000 g of the finely ground, dried (110°C) sample into a platinum crucible. Place the crucible in a muffle furnace and slowly raise the temperature to $1000 \pm 25°C$. Ignite to constant weight at this temperature; 30 min ignition is usually sufficient.

For routine work it is possible to start the ignition over a low mushroom flame, slowly increasing the temperature to full heat over a period of about 20 min after which the crucible is transferred to a furnace at 1000°C for 30 min.

Determination of Silica

NOTE: The determination should be carried out from the completion of the fusion to the filtration of the precipitate without delay.

Decomposition of the Sample

Weigh 0·100 g of the finely ground, dried (110°C) sample into a platinum dish; a convenient size is 70-mm dia., 40-mm deep, of effective capacity 75 ml.

Add 4 g of fusion mixture and 2 g of boric acid and mix intimately to form a charge in the centre of the dish.

Heat over a Bunsen or Méker flame, slowly raising the temperature until the mixture begins to melt and keep at this temperature until melting is complete. It is important that the initial stage of heating should be slow, since a vigorous reaction takes place on melting, and spattering easily occurs. Again raise the temperature slowly and steadily to that of the full heat of the flame (approx. 950°C). After 5 min at this temperature, swirl the contents of the dish every 2 min making sure that the particles of sample on the side of the dish come into contact with hot, molten flux. Continue heating at this temperature until decomposition is complete, 15 min is usually sufficient. This can be checked visually by the absence of unfused material at the bottom of the melt. The type and position of the burner should be chosen so as to maintain oxidizing conditions throughout.

Cool, place the dish and lid in a porcelain basin so as to avoid loss of solution if frothing occurs and add 20 ml of diluted sulphuric acid (1 + 3). When dissolution is complete transfer the solution to a 1000-ml beaker, washing the dish and lid with water. Dilute the solution to approx. 400 ml with cold water.

Determination of Silica

Immediately add 50 ml of ammonium molybdate solution (100 g/litre) from a pipette or burette, stirring vigorously during the addition and for 1 min after it, and stand at room temperature for 10 min.

Add 50 ml of hydrochloric acid ($d = 1\cdot18$) rapidly, and immediately precipitate the quinoline silicomolybdate by adding 50 ml of quinoline solution (2 % v/v) from a burette, dropwise at first but rapidly after the formation of a permanent precipitate, stirring the solution during the addition.

Place a thermometer in the solution and heat with occasional stirring to 80°C over a period of about 10 min and maintain at this temperature for 5 min. Cool, stirring occasionally, in running water to *less than 20°C* and filter through a weighed Gooch crucible with a sintered-glass mat (porosity 4) transferring the precipitate to the crucible with a jet of quinoline wash solution (0·05 % v/v) also cooled to less than 20°C. After scrubbing out the beaker with a "bobby", wash the precipitate eight times with quinoline wash solution, CARE BEING TAKEN NEVER TO LET THE PRECIPITATE RUN DRY DURING FILTRATION AND WASHING.

Dry the precipitate at 150°C to constant weight; 2 h is usually sufficient. Cool in a desiccator and weigh.

Calculation

Weight of precipitate (g) $\times$ 25·66 = % SiO_2 (if 0·1-g sample taken). Correct for the blank determination.

Determination of Chromium Sesquioxide

Decomposition of the Sample

Weigh 0·250 g of the finely ground, dried (110°C) sample into a platinum dish; a convenient size is 70-mm dia. 40-mm deep, of effective capacity 75 ml.

Add 4 g of fusion mixture and 2 g of boric acid and mix intimately to form a charge in the centre of the dish.

Heat over a Bunsen or Méker flame, slowly raising the temperature until the mixture begins to melt and keep at this temperature until melting is complete. It is important that the initial stage of heating should be slow, as a vigorous reaction takes place on melting, and spattering easily occurs. Again raise the temperature slowly and steadily to that of the full heat of the flame (approx. 950°C). After 5 min at this temperature, swirl the contents of the dish every 2 min making sure that the particles of sample on the sides of the dish come into contact with hot, molten flux. Continue heating at this temperature until decomposition is complete, 15 min is usually sufficient. This can be checked visually by the absence of unfused material at the bottom of the melt. The type and position of the burner should be chosen so as to maintain oxidizing conditions throughout.

Cool, place the dish and lid in a porcelain basin so as to avoid loss of solution if frothing occurs and add 20 ml of diluted sulphuric acid (1 + 3). When decomposition is complete, transfer the solution to a 400-ml beaker and add 100 ml of diluted sulphuric acid (1 + 3).

Determination of Chromium Sesquioxide

Transfer the solution to a 1-litre conical flask and dilute to about 500 ml. Heat to boiling and add potassium permanganate solution (approx. $0.1N$) slowly, drop by drop, until the reddish-amber colour of the solution persists after boiling for 5 min. Normally, less than 2 ml is required and it is important that no great excess of permanganate be added.

Cool slightly, add 20 ml of diluted hydrochloric acid (1 + 3), destroy the excess of permanganate and any manganese dioxide by boiling until the solution is crystal clear, then boil for a further 15 min to remove all traces of chlorine.

Cool, add 25 ml, 40 ml or 50 ml as necessary, of the ferrous ammonium sulphate solution (approx. $0.1N$)* from a pipette and then add 25 ml of phosphoric acid ($d = 1.75$) and 10 drops of barium diphenylamine sulphonate indicator. Back-titrate the excess ferrous ammonium sulphate with standard potassium dichromate solution ($0.05N$).

Standardize the ferrous ammonium sulphate against the standard potassium dichromate as follows:

Dilute 120 ml of sulphuric acid (1 + 3) to about 500 ml with water in a 1-litre conical flask. Add 25 ml, 40 ml or 50 ml according to the volume taken in the actual determination, of the ferrous ammonium sulphate solution (approx. $0.1N$)* using the same pipette as in the determination. Add 25 ml of phosphoric acid ($d = 1.75$) and 10 drops of barium diphenylamine sulphonate indicator and titrate with the standard potassium dichromate solution ($0.05N$).

Calculation—If V is the volume of potassium dichromate solution ($0.05N$) equivalent to the ferrous ammonium sulphate added and v ml the volume used in the back-titration, then the percentage of Cr_2O_3 is $0.5067 (V-v)$ if exactly 0.25 g of the sample has been taken.

*It has been found adequate, provided that reasonable precautions are taken to ensure homogeneity, to use a weighed amount of solid ferrous ammonium sulphate; standardization is then made on a weight of the salt similar to that used in the determination.

Determination of Titania, Ferric Oxide, Alumina, Manganese Oxide, Lime and Magnesia

Decomposition of the Sample

Weigh 1·000 g of the finely ground, dried (110°C) sample into a platinum dish. Add 8 g of fusion mixture and 4 g of boric acid and mix thoroughly. Cover the dish with a lid.

Heat over a Bunsen or Méker flame, slowly raising the temperature until the mixture begins to melt, and keep at this temperature until melting is complete. It is important that the initial stage of heating should be slow, as a vigorous reaction takes place on melting, and spattering easily occurs. Again raise the temperature slowly and steadily to the full heat of the flame (approx. 950°C).

After about 5 min at this temperature, swirl the contents of the dish every 2 min making sure that the particles of sample on the sides of the dish come into contact with hot, molten flux.

If swirling is begun at too early a stage, it is difficult to detach unfused particles from the sides of the dish.

Continue heating at this temperature for 45–60 min depending on the nature of the sample. Completeness of decomposition can be checked visually by the absence of any black, unfused material at the bottom of the melt. The type and position of the burner should be chosen so as to maintain oxidizing conditions throughout.

Cool, place the dish in a porcelain basin of about 5-in. dia. (This is to avoid loss of solution if the reaction on the addition of the acid is too vigorous.) Raise the lid slightly and add 35 ml of diluted sulphuric acid (1 + 3). Replace the lid and allow the reaction to proceed for a few minutes and, if necessary, complete the dissolution by warming. Transfer the solution from the dish to a 400-ml beaker, dilute to approx. 100 ml and cool.

Extraction of Chromate

NOTE: The separation should be carried out in diffuse light, not bright sunlight, to minimize reduction of chromate.

Transfer the cold solution to a 500-ml separating funnel (Squibb's type) with the minimum of water. The volume at this stage should not exceed 150 ml. Add 50 ml of Amberlite LA-2 resin solution (1 + 4). Stopper the funnel and shake vigorously for 1 min. Release the pressure in the funnel by carefully removing the stopper; rinse the stopper and neck of the funnel with water. Allow the layers to separate, then withdraw and discard the organic phase. Repeat the extraction with 20 ml of the resin solution (1 + 4) and again withdraw the organic phase.

Add 10 ml of diluted sulphuric acid (1 + 9) to the aqueous phase and repeat the extraction with a further 20 ml of resin solution (1 + 4), again discarding the organic phase.

Remove the traces of resin by two successive extractions with 20-ml portions of chloroform. Transfer the aqueous phase and washings from the funnel to a 400-ml beaker and boil the solution, to remove traces of chloroform. Cool, dilute to 500 ml in a volumetric flask and mix. This solution is referred to as the "stock solution".

Determination of Titania

Transfer a 20-ml aliquot of the stock solution to each of two 50-ml volumetric flasks, A and B. To each flask add 10 ml of diluted phosphoric acid $(2 + 3)$ and, to flask A only, 10 ml of hydrogen peroxide solution (6%). Dilute the solution in each flask to 50 ml and shake well.

Measure A against B in 40-mm cells at 398 nm, or by using a colour filter (Ilford 601) in a suitable instrument. The colour is stable from 5 min until 24 h after the addition of the hydrogen peroxide solution.

Determine the titania content of the solution by reference to a calibration graph.

Determination of Total Iron (as Ferric Oxide)

NOTE: The dilution of the stock solution quoted will cover the range 0–20% Fe_2O_3. For iron contents considerably below 20% Fe_2O_3 a decreased dilution of the stock solution should be made. For contents above 20% Fe_2O_3 an increased dilution of the stock solution should be made. It is not permissible to dilute an aliquot of the solution once the colour has developed.

Dilute 25·0 ml of the stock solution to 100 ml in a volumetric flask and mix. Transfer 5·0 ml of this diluted solution to a 100-ml volumetric flask and add 2 ml of hydroxyammonium chloride solution (100 g/litre) followed by 5 ml of 1:10-phenanthroline solution (10 g/litre). Add ammonium acetate solution (approx. 10%) until a pink colour forms then add 2 ml in excess. Allow to stand for 15 min, dilute the solution to 100 ml and mix.

Measure the optical density of the solution against water in 10-mm cells at 510 nm or by using a colour filter (Ilford 603) in a suitable instrument. The colour is stable from 15 min to 75 min after the addition of the ammonium acetate solution. Determine the ferric oxide content of the solution by reference to a calibration graph.

Determination of Manganese Oxide

Transfer a 50-ml aliquot of the stock solution to a 250-ml beaker. Add 10 ml of diluted sulphuric acid $(1 + 1)$, 5 ml of nitric acid $(d = 1·42)$ and evaporate to strong fumes to destroy traces of resin.

Cool, add 20 ml of nitric acid $(d = 1·42)$, 10 ml of diluted phosphoric acid $(1 + 9)$ and about 50 ml of water.

Boil to dissolve salts and to remove nitrous fumes, filtering if necessary. Add about 0·5 g of potassium periodate, boil for 2 min and transfer to a steam bath for 10 min.

Allow to cool and transfer to a 100-ml volumetric flask. Dilute the solution to 100 ml and mix. Measure the optical density of the solution against water in 40-mm cells at 524 nm or by using a colour filter (Ilford 604) in a suitable instrument.

Determine the manganese oxide content of the solution by reference to a calibration graph.

Determination of Alumina

Transfer a 100-ml aliquot of the stock solution to a 500-ml separating funnel, Squibb's type. Add ammonia solution ($d = 0.88$) dropwise, until the solution is faintly alkaline to bromophenol blue. Just reacidify with diluted hydrochloric acid ($1 + 3$) and add 4 ml in excess.

Add 20 ml of chloroform and 10 ml of sodium diethyldithiocarbamate solution (100 g/litre). Stopper the funnel and shake vigorously. Release the pressure in the funnel by carefully removing the stopper and rinse the stopper and neck of the funnel with water. Allow the layers to separate and withdraw the chloroform layer (if an emulsion has formed it will be necessary to add a few drops of hydrochloric acid and reshake).

Add 10-ml portions of chloroform and repeat the extraction with 5-ml portions of DDC until a coloured precipitate is no longer formed. Finally wash the aqueous phase with 20 ml of chloroform. This separation will remove iron and manganese.

Add 25 ml of hydrochloric acid ($d = 1.18$) followed by 2–3 ml of cupferron solution (60 g/litre) and 20 ml of chloroform. Stopper the funnel and shake vigorously. Remove the stopper and rinse the stopper and neck of the funnel with water. Allow the layers to separate and withdraw the chloroform layer. Repeat the extraction with three 10-ml portions of chloroform to remove traces of cupferron and DDC.

Run the aqueous phase from the separating funnel into a 1-litre conical flask. Add a few drops of bromophenol blue indicator followed by ammonia solution ($d = 0.88$) until just alkaline.

Reacidify quickly with hydrochloric acid, add 5–6 drops in excess and cool the flask in running water. The solution must be cold before addition of DCTA or the residual chromium present in the stock solution may be complexed.

Add sufficient standard DCTA solution (0.05M) approx. to produce an excess of a few millilitres over the expected amount (1 ml ≡ 1.275% Al$_2$O$_3$). Then add ammonium acetate buffer solution (for Al$_2$O$_3$ determination) until the indicator turns blue followed by 15 ml in excess.

Add a volume of ethanol (95%) equal to the total volume of the solution, 20 ml of hydroxyammonium chloride solution (100 g/litre) and 1–2 ml of dithizone indicator. The hydroxyammonium chloride solution is added before the dithizone indicator and prevents the latter from being decolorized by the traces of cupferron present in the solution. This cupferron is not destroyed as the normal boiling stage is eliminated. Titrate with standard zinc solution (0.05M) from green to the first appearance of a permanent pink colour. (The end-point is often improved by the addition of a little Naphthol Green B solution (1 g/litre) so as to eliminate any pink colour which may be formed in the solution on the addition of the indicator.)

Calculation—If the DCTA solution is not exactly 0.05M, calculate the equivalent volume of 0.05M DCTA.

If V ml is the volume of DCTA (0.05M) added and v ml is the volume of standard zinc solution (0.05M) used in the back-titration then

$$Al_2O_3\% = 1.275\,(V - v)$$

Determination of Lime

Transfer a 100-ml aliquot of the stock solution to a 250-ml volumetric flask. Add 5 ml of diluted triethanolamine (1 + 1), 5·0 ml of standard EGTA solution (0·05M approx.) and dilute to 150 ml. Add potassium hydroxide solution (250 g/litre) until no further precipitation takes place, add 10 ml in excess followed by 10 ml of Magflok solution (20 g/litre). Dilute to 250 ml, shake and allow to stand for 10 min to settle. Filter through a 150-mm *dry*, coarse filter paper (Whatman No. 541) into a *dry* beaker. Pipette 200 ml of the filtrate into a 500-ml conical flask and add 15 ml of potassium hydroxide solution (250 g/litre). Add approx. 0·03 g of screened Calcein indicator and titrate with standard lime solution (1·0 mg CaO/ml) until the first appearance of a green fluorescence.

The titration is best carried out in good daylight, but direct sunlight should be avoided.

Calculation—If 1 ml of EGTA $\equiv x$ mg CaO and y ml are added initially and t ml is the back-titration with standard lime solution (1 mg CaO/ml) then

$$\text{CaO} \% = 0·625 \left(\tfrac{4}{5} xy - t\right)$$

Correct for the blank determination.

Determination of Magnesia

Transfer 100·0 ml of the stock solution to a 500-ml separating funnel and add ammonia solution ($d = 0·88$) dropwise until the solution is faintly alkaline to bromophenol blue. Just reacidify with diluted hydrochloric acid (1 + 3) and add 4 ml in excess.

Add 20 ml of chloroform and 10 ml of sodium diethyldithiocarbamate solution (100 g/litre). Stopper the funnel and shake vigorously. Release the pressure in the funnel by carefully removing the stopper and rinse the stopper and neck of the funnel with water. Allow the layers to separate and withdraw the chloroform layer. (If an emulsion has formed it will be necessary to add a few drops of hydrochloric acid and reshake.)

Add 10-ml portions of chloroform and repeat the extraction with 5-ml portions of DDC until a *coloured* precipitate is no longer formed. Finally wash the aqueous phase once with 10 ml of chloroform. This separation will remove iron and manganese.

Add diluted ammonia solution (1 + 1) dropwise until the solution is just alkaline to bromophenol blue. *Just* reacidify with diluted hydrochloric acid (1 + 9) and add 20 ml of the ammonium acetate buffer solution (for magnesia determination).

Add 20 ml of chloroform and 10 ml of cupferron solution (60 g/litre). Stopper the funnel and shake vigorously. Release the pressure in the funnel by carefully removing the stopper and rinse the stopper and neck of the funnel with water. Allow the layers to separate and withdraw the chloroform layer. Repeat the extraction with a further 10 ml of cupferron solution and finally wash the aqueous phase three times with 10-ml portions of chloroform. This separation will remove aluminium and titanium.

Transfer the aqueous layer to a 500-ml conical flask and boil off traces of chloroform. Cool and add 2 g of ammonium chloride and 5 ml of diluted triethanolamine (1 + 1) with swirling followed by an appropriate known amount of DCTA solution (approx. 0·05M). This addition is made to complex most of the magnesia before the solution is made alkaline so that the tendency for magnesium hydroxide to precipitate is greatly reduced. Then add 30 ml of ammonia solution ($d = 0·88$) and 5 ml of hydroxyammonium chloride solution (100 g/litre) to stabilize the indicator.

Titrate with the standard DCTA solution using Solochrome Black 6B indicator, from red through purple to the last change to a clear blue. This titration also includes the titration for lime in the sample which must be allowed for.

Calculation—If the DCTA solution is not exactly 0·05M, calculate the equivalent volume of exactly 0·05M DCTA solution.

If V is the total amount of 0·05M DCTA added and v is the amount of 0·05M DCTA required to react with the lime in 0·2-g sample, then % MgO = 1·008 $(V - v)$. Correct for the blank determination.

Determination of Alkalis

Determine the alkalis in the sample by the procedure given in Chapter 16 under "Chrome-bearing Materials".

CHAPTER 13

ANALYSIS OF ZIRCON AND ZIRCON-BEARING REFRACTORIES

GENERAL

The need to use the classical method for the analysis of zircon and zircon refractories has now largely disappeared. The application of coagulation to the determination of silica followed by the removal of the zirconia by precipitation as the cupferrate can be made to yield a solution on which the determination of all other relevant constituents with the exception of alkalis can be completed.

Other methods for the determination of zirconium are more specific than precipitation with cupferron, but the filtrates are not so amenable to treatment to allow further determinations. The iron and titanium which are precipitated with the zirconium are small in amount and the determination of zirconia by difference appears well justified. Another relevant point concerning the procedure is that the silica not removed by coagulation does not precipitate with the cupferrates but remains in solution and can be determined colorimetrically.

PRINCIPLE OF THE METHOD

The loss on ignition is determined by heating the sample to 1000°C for 30 min.

The sample is fused in alkali carbonate and boric acid in a platinum dish and the melt dissolved in hydrochloric and sulphuric acids. A gel of silica is formed by briefly evaporating and this is coagulated with polyethylene oxide. The coagulation avoids the hydrolysis of a large proportion of the zirconium, such as occurs on dehydration, and the silica precipitate filters much more readily. After washing, the silica precipitate is ignited and weighed, before and after treatment with hydrofluoric acid. The small residue after the hydrofluoric acid treatment is fused in potassium pyrosulphate, the melt dissolved in acid and combined with the filtrate from the silica determination.

After adjustment of the acidity, cupferron is added to precipitate zirconium, iron and titanium in the combined filtrate, and the precipitate is filtered off, washed, carefully dried and ignited and finally weighed as the mixed oxides. The filtrate is shaken with chloroform to remove the green cupferron coloration and then boiled with hydrogen peroxide. This destroys the decomposition products which would normally result in a brown precipitate when the solution was made alkaline, and thus permits the determination of lime, magnesia and alumina by EDTA. The solution is then diluted to a known volume and aliquots taken for further determinations. Residual silica is determined spectrophotometrically as the yellow silicomolybdate. Lime is determined by the addition of excess EGTA and back-titration with standard lime solution using Calcein as indicator; end points are much sharper by this technique than by the more usual titration with EDTA. Magnesia plus lime is determined by direct titration with EDTA using Solochrome Black 6B as indicator and then magnesia is obtained by difference. Finally, alumina is determined by adding excess EDTA and back-titration with standard zinc solution using dithizone as indicator.

The mixed oxides of zirconium, iron and titanium are fused in potassium pyrosulphate and the melt dissolved in diluted sulphuric acid. After diluting to volume, aliquots are used for the spectrophotometric determination of iron and titanium with 1:10-phenanthroline and hydrogen peroxide respectively. Zirconium is then obtained by calculation.

Alkalis are determined on a solution made by decomposing the sample with hydrofluoric, sulphuric and nitric acids, evaporating, dissolving the residue in nitric acid, filtering and making up to volume. The determination is carried out using a flame-photometer.

REAGENTS

Prepared Reagents

Ammonium acetate (approx. 10%): Dilute 140 ml of acetic acid (glacial) to 2000 ml with water and add carefully 140 ml of ammonia solution ($d = 0.88$). Mix, cool and adjust to approximately pH 6.0 either with acetic acid or ammonia solution.

Ammonium acetate buffer: Add 120 ml of acetic acid (glacial) to 500 ml of water, followed by 74 ml of ammonia solution ($d = 0.88$). Mix, cool, and dilute to 1 litre.

Ammonium molybdate (80 g/litre): Dissolve 80 g of ammonium molybdate in 1 litre of water; filter if necessary. Store in a polythene bottle and discard after 4 weeks, or earlier if any appreciable deposit of molybdic acid is observed.

Cupferron (60 g/litre): Dissolve 3 g of cupferron in 50 ml of cold water and filter. This solution must be freshly prepared. If the reagent is discoloured or gives a strongly coloured solution a new stock should be obtained. The solid reagent should be stored in a tightly-stoppered bottle in the presence of a piece of ammonium carbonate to prevent decomposition.

Cupferron wash liquor: Add 100 ml of hydrochloric acid ($d = 1.18$) to 800 ml of water. Dissolve 1.5 g of cupferron in 100 ml of cold water and filter. Mix the two solutions. This wash liquor should be freshly prepared.

1:10-Phenanthroline (10 g/litre): Prepare enough solution for immediate use at a concentration of 0.1 g of 1:10-phenanthroline hydrate in 10 ml of diluted acetic acid (1 + 1).

Polyethylene oxide (2.5 g/litre): Add 0.5 g of polyethylene oxide* to 200 ml of water slowly with stirring, preferably on a mechanical stirrer, until dissolved. Discard after 2 weeks.

* Union Carbide Polyox resins WSR 35, WSR-N-80, WSR 205, WSR-N-750 and WSR-N-3000 are suitable as a source of polyethylene oxide.

Indicators

Bromophenol blue
Calcein (screened)
2,4-Dinitrophenol
Dithizone
Solochrome Black 6B

Standard Solutions

Calcium (1 mg CaO/ml)
EDTA (0·05M)
EDTA (5 g/litre)
EGTA (5 g/litre)
Zinc (0·05M).

PREPARATION OF SAMPLE

The sample prepared for analysis should be ground to pass completely through a 200-mesh B.S. test sieve. A non-metallic (e.g. nylon bolting-cloth) sieve is preferable.

Zirconia-bearing materials are often extremely hard and sample preparation is difficult. For routine work it may be sufficient to crush the sample in an iron percussion mortar and then treat with a magnet. For more accurate work it will probably be necessary to prepare a separate sample for an iron determination in an alumina or possibly an agate mortar. The main analysis should then be performed on the iron-ground sample and an iron determination on the other sample. The results should then be calculated to the latter iron content. Any metallic iron introduced will be oxidized to the ferric state by the heat treatment during the determination of the loss on ignition. Correction must be made for the difference in ferric oxide content of the two samples, expressed as metallic iron, the state in which it is present when weighing out the sample.

The use of a boron carbide mortar permits the grinding of zircon materials with the minimum of contamination, thus obviating the necessity for preparing two separate portions of the sample.

PROCEDURE

Determination of Loss on Ignition

Weigh 1·000 g of the finely ground, dried (110°C) sample into a platinum crucible. Place the crucible in a muffle furnace and slowly raise the temperature to 1000 ± 25°C. Ignite to constant weight at this temperature; 30 min ignition is usually sufficient.

For routine work, it is possible to start the ignition over a low mushroom flame, slowly increasing the temperature to full heat over a period of about 20 min after which the crucible is transferred to a furnace at 1000°C for 30 min.

Determination of Silica, Alumina, Ferric Oxide, Titania, Zirconia, Lime and Magnesia

Decomposition of the Sample

Weigh 1·000 g of the finely ground, dried (110°C) sample into a platinum basin: a convenient size is 70-mm dia., 40-mm deep, of effective capacity 75 ml. Add 5 g of fusion mixture and 2 g of boric acid and mix intimately.

For routine work, the loss on ignition may be carried out in the above dish and the same portion of sample used for the fusion. This technique is only admissible if the amount of sintering of the sample after the determination of loss on ignition does not prevent adequate mixing of the sample and flux.

Heat over a mushroom burner, cautiously at first, then gradually raise the temperature to the full heat of the burner over a period of about 10 min, the dish being covered with a lid to prevent loss by spurting. In some cases it may be necessary transfer to the dish to a Méker burner to melt the flux effectively.

Finally heat the dish and contents in a muffle furnace at 1200°C for 15 min. Remove the dish from the furnace and allow to cool.

Add 15 ml of diluted sulphuric acid (1 + 1) and 20 ml of diluted hydrochloric acid (1 + 1) and transfer the dish to a steam bath to facilitate dissolution of the melt. If the solution gels before dissolution is complete, it may be necessary to stir the gel gently with a glass rod. When solution is complete, remove the lid and wash any spray adhering to it into the dish with the minimum amount of water.

Determination of Gravimetric Silica

Allow the dish to remain on the steam bath until a stiff gel is formed and thoroughly mix a Whatman accelerator tablet into the gel until the mixture is homogeneous, ensuring that the gel adhering to the sides of the dish is also broken up.

Add slowly, with stirring, 5 ml of polyethylene oxide coagulant solution (2·5 g/litre) and mix well, ensuring that any gel adhering to the sides of the dish is brought into contact with the coagulant. Add 10 ml of water, mix and allow to stand for 5 min.

Filter through a 110-mm No. 42 Whatman paper, transferring the silica to the filter with hot, diluted hydrochloric acid (1 + 19), scrubbing the dish with a rubber-tipped glass rod. Wash the precipitate five times with hot, diluted hydrochloric acid (1 + 19) and then with hot water until free from chlorides (usually to a volume of about 400 ml). The silica precipitate obtained by this method is more voluminous than that obtained by dehydration and so care must be taken to ensure that the mass of precipitate is broken up thoroughly during the washing. Reserve the filtrate and washings.

Transfer the paper and precipitate to an ignited and weighed platinum crucible. Ignite at a low temperature until the precipitate is free from carbonaceous matter and then heat in a muffle furnace at 1200°C to constant weight, 30 min usually being sufficient. (The ignition should be started at a lower temperature than that normally used or a noticeable amount of carbon may remain after ignition. In some cases a trace of carbon remains but its weight appears to be negligible and does not affect the results.)

Moisten the contents of the cold crucible with water, add 5 drops of diluted sulphuric acid (1 + 1) and 10 ml of hydrofluoric acid (40 % w/w). Evaporate to dryness on a sand bath in a fume cupboard.

For the evaporation, the crucible and contents should be heated from below; the use of top heating alone, as with a radiant heater, can result in incomplete elimination of silica by hydrofluoric acid.

Heat the crucible and residue, cautiously at first, over a gas flame and finally for 5 min at 1200°C. Cool and weigh. The difference between the two weights represents the "gravimetric" silica.

Preparation of the Combined Solution of the Silica Filtrate and Silica Residue

To the filtrate from the determination of main silica add 40 ml of hydrochloric acid ($d = 1\cdot18$) and boil the solution down to a volume of about 250 ml. Prolonged boiling should be avoided.

Fuse the residue from the hydrofluoric acid treatment of the "gravimetric" silica with about 2 g of potassium pyrosulphate. Dissolve the melt in 10 ml of water and 5 ml of diluted sulphuric acid (1 + 1), in the crucible, over a low gas flame (the "pilot" jet of a bunsen burner is suitable). Add this solution to the boiled-down silica filtrate to form the "combined solution".

Determination of the Sum of the Mixed Oxides of Zirconium, Titanium and Iron

Cool the "combined solution" to a temperature of less than 10°C in icewater, or preferably a refrigerator.

Add 125 ml of cold cupferron solution (60 g/litre), slowly with stirring, followed by a small amount of filter-paper pulp (ashless). Stand the beaker and contents in the cold for 5 min and add a few more drops of cupferron solution (60 g/litre) to check that no further precipitation takes place.

Filter through a 150-mm No. 42 Whatman paper, transferring the precipitate to the filter with a jet of cold cupferron wash liquor and scrubbing the beaker with a "bobby". Wash the precipitate thoroughly with cold cupferron wash liquor.

It is essential to keep the solution cold during filtration and washing, otherwise the precipitate tends to become tarry and cannot be effectively washed.

Transfer the precipitate to an ignited and weighed platinum *dish*, 70-mm dia. Reserve the filtrate and washings; if further analysis is required then this solution should be treated as given on p. 141 as soon as possible.

Place the dish and contents in an oven at 110°C until the precipitate is dry and then burn off the carbon at a low temperature in a fume cupboard, great care must be taken at this stage as the precipitate itself tends to melt and some may be lost by spurting. Finally, ignite the mixed oxides at 1200°C to constant weight; 30 min is usually sufficient. Cool and weigh. The weight is that of zirconia, titania and ferric oxide in the 1-g sample.

Carefully fuse the ignited, weighed residue in the same platinum basin with 15 g of potassium pyrosulphate. Cool the dish and dissolve the melt in 30 m of water and 20 ml of diluted sulphuric acid (1 + 1), in the dish, over a low flame (the "pilot" jet of a bunsen burner is suitable).

Cool the solution, dilute to 250 ml in a volumetric flask and mix.

Determination of Ferric Oxide

Transfer a 5-ml aliquot of the solution of the ignited cupferron precipitates to a 200-ml volumetric flask. (A larger aliquot may be used if greater accuracy is desired.) Add 2 ml of hydroxyammonium chloride solution (100 g/litre) and 5 ml of 1:10-phenanthroline solution (10 g/litre). Add ammonium acetate solution (approx. 10%) until a pink colour develops in the solution and then add 2 ml in excess.

Allow to stand for 15 min, dilute to 200 ml and mix.

Measure the optical density of the solution against water in 10-mm cells at 510 nm or by using a colour filter (Ilford 603) in a suitable instrument. The colour is stable between 15 min and 75 min after the addition of the ammonium acetate solution.

Determine the ferric oxide content of the solution by reference to a calibration graph.

Determination of Titania

Transfer a 10-ml aliquot of the solution of the ignited cupferron precipitates to each of two 50-ml volumetric flasks, A and B. (A larger aliquot may be used if greater accuracy is desired.) Add to flask A only, 10 ml of hydrogen peroxide solution (6%), dilute the solution in each flask to 50 ml and shake well.

Measure A against B in 40-mm cells at 398 nm or by using a colour filter (Ilford 601) in a suitable instrument. The colour is stable between 5 min and 24 h after the addition of the hydrogen peroxide solution. Determine the titania content of the solution by reference to a calibration graph (prepared without the addition of phosphoric acid).

Calculation of Zirconia

The weight of the ignited cupferron precipitate, less the sum of the weights of the ferric oxide and the titania in the 1-g sample, is the weight of the zirconia in the 1-g sample.

It should be noted that this method does not distinguish between zirconia and hafnia. Thus the figure obtained for zirconia is actually $ZrO_2 + HfO_2$.

Preparation of the Solution for the Determination of Residual Silica, Alumina, Lime and Magnesia

Transfer the filtrate and washings, reserved from the cupferron precipitation, to a 1000-ml separating funnel. Add 20 ml of chloroform and shake vigorously. Release the pressure in the funnel by carefully removing the stopper and rinse the stopper and neck of the funnel with water.

Allow the layers to separate and withdraw the chloroform layer. Repeat the extraction with two further 20-ml portions of chloroform and discard the chloroform extracts.

Transfer the aqueous solution from the extraction to a 1000-ml beaker and add 100 ml of hydrogen peroxide solution (6%). Boil the solution until the volume is about 400 ml; at this stage the solution should be colourless or slightly greenish, if the solution is unduly coloured then a further 25 ml of hydrogen peroxide should be added and the solution boiled for 15 min.

Allow the solution to cool, dilute to 500 ml in a volumetric flask and mix. The solution thus obtained will be referred to as the "stock test solution".

Determination of Residual Silica

Transfer a 20-ml aliquot of the "stock test solution" to each of two 50-ml volumetric flasks, A and B, and add a few drops of 2 : 4-dinitrophenol indicator. To flask B add ammonia solution ($d = 0.88$) dropwise until the indicator turns yellow (note the number of drops used) then add 5 ml of diluted hydrochloric acid (1 + 4). To flask A add 5 ml of diluted hydrochloric acid (1 + 4) and the same number of drops of ammonia as needed to neutralize the solution in flask B, then add 5 ml of ammonium molybdate solution (80 g/litre).

Dilute the solution in each flask to 50 ml and shake well. Measure the silicomolybdate colour in flask A against the solution in flask B in 40-mm cells at 440 nm or by using a colour filter (Ilford 601) in a suitable instrument.

Measure the optical density not sooner than 5 min and not later than 15 min after the addition of the ammonium molybdate solution.

Determine the silica content of the solution by reference to a calibration graph.

Calculation of Total Silica Content

Add the residual silica content to the figure obtained for the "gravimetric" silica to obtain the total silica content.

Determination of Alumina

The method described below is for normal zircon-bearing refractories where the lime content is low. If lime has been deliberately added then about 10 g of ammonium sulphate should be added to the aliquot taken for the alumina determination.

Transfer a 200-ml aliquot of the "stock test solution" to a 500-ml conical flask. (If the alumina content is known to be high then the normal 100-ml aliquot may be used.) Add a few drops of bromophenol blue indicator and add ammonia solution ($d = 0.88$) until just alkaline.

Re-acidify quickly with hydrochloric acid ($d = 1.18$) and add 5–6 drops in excess.

Add sufficient EDTA solution (0.05M) to provide an excess of a few millilitres over the expected amount. Then add ammonium acetate buffer solution until the solution turns blue, followed by 10 ml in excess. Boil the solution for 10 min and cool rapidly.

Add an equal volume of ethanol (95%) and 1–2 ml of dithizone indicator, and titrate with standard zinc solution (0.05M) from green to the first appearance of a permanent pink colour.

Calculation—If the EDTA solution is not exactly 0·05M, calculate the equivalent volume of 0·05M EDTA.

If V ml is the volume of EDTA (0·05M) and v is the volume of standard zinc solution (0·05M) used in the back-titration, then

percentage of Al_2O_3 = 1·275 $(V - v)$ for a 100-ml aliquot and
percentage of Al_2O_3 = 0·6375 $(V - v)$ for a 200-ml aliquot.

Determination of Lime

Transfer a 50-ml aliquot of the "stock test solution" into a 500-ml conical flask. Add 5 ml of diluted triethanolamine (1 + 1), 5·00 ml of standard EGTA solution (5 g/litre), 50 ml of potassium hydroxide solution (250 g/litre) and dilute the solution to about 200 ml.

Add about 0·015 g of screened Calcein indicator and titrate with standard lime solution (1 mg CaO/ml) from a semi-micro burette from red to the first appearance of a green fluorescence.

Calculation—If a 0·1-g aliquot is taken and 1 ml of the EGTA solution $\equiv x$ mg CaO and V ml of standard lime solution (1 mg CaO/ml) are used in the back titration, then the percentage of CaO = $5 x - V$

NOTE: The volume of EGTA added is satisfactory for up to 3% CaO. For samples containing a higher CaO content an appropriate amount of EGTA should be added.

Determination of the Sum of Lime and Magnesia

Transfer a 50-ml aliquot of the "stock test solution" to a 500-ml conical flask. Add 20 ml of diluted triethanolamine (1 + 1), 40 ml of ammonia solution (d = 0·88) and 5 ml of hydroxyammonium chloride solution (100 g/litre) and dilute to 200 ml.

Add Solochrome Black 6B indicator and titrate with standard EDTA solution (5 g/litre) from a semi-micro burette from red through purple to the last change to a clear ice-blue.

Calculation of Magnesia

Calculate the milligrammes of CaO in 0·1-g sample (m).

If 1 ml of EDTA $\equiv y$ mg CaO then the volume of EDTA used to titrate the lime = m/y.

If V = total amount of EDTA to titrate the CaO + MgO then the volume of EDTA to titrate the MgO = $V - m/y$ ml.

Determination of Alkalis

Determine the alkalis in the sample by the procedure given in Chapter 16 under "Zircon-bearing materials".

CHAPTER 14

ANALYSIS OF BONE ASHES

GENERAL

Bone ash and bone china body are not materials which are frequently analysed in the ceramic industries. Thus, there has been no demand for the development of more rapid methods, so that although the present method is slow and tedious it is the only procedure which can be recommended for accurate work. It is based on that of Harvey: "Scheme for the Complete Analysis of Apatite Rocks" (*Analyst* **61**, 817, 1936)

PRINCIPLE OF THE METHOD

The loss on ignition is determined by heating the sample to 1000°C for 30 min.

The silica is determined on a separate sample by dissolving material soluble in hydrochloric acid, filtering and fusing the residue. The melt is dissolved in hydrochloric acid and the combined solutions evaporated to dryness. The insoluble silica is filtered off and determined by ignition and weighing before and after hydrofluoric acid treatment. The silica remaining in solution is determined by the molybdenum blue procedure in which phosphate does not interfere.

The sample used for the determination of alumina, ferric oxide, titania, lime and magnesia is first dissolved in hydrochloric acid, and the residue treated with hydrofluoric and sulphuric acids and the solution evaporated to strong fuming. The solution of the residue and the original solution are combined in a platinum dish, sulphuric acid added and combined solutions evaporated to strong fumes. Lime is now precipitated as the sulphate by the addition of water and ethanol. The filtrate is reserved for the determination of R_2O_3 and magnesia, the sulphate precipitate is dissolved in hydrochloric acid, and any residue dissolved by fusion. An ammonia precipitation ensures the freedom of the solution from the R_2O_3 group, any precipitate being dissolved and added to the reserved filtrate.

A double precipitation with oxalate, followed by ignition of the second precipitate yields a figure for the CaO content. Any magnesia which may have been co-precipitated with the $CaSO_4$ is precipitated in the filtrates from the lime as the phosphate, filtered, re-dissolved and added to the main filtrate.

The R_2O_3 group is now precipitated as the phosphate, if the precipitate is large a second precipitation is needed. The precipitate is ignited and then normally fused in potassium pyrosulphate. Titania is determined on the solution by hydrogen peroxide and iron by 1:10-phenanthroline, and the alumina content derived by calculation. The presence of phosphate in the precipitate will result in low figures for ferric oxide but the iron content is usually not needed exactly and is, in any event, very low. If an accurate answer is required the ignited precipitate may be first fused with sodium carbonate, and the sodium phosphate filtered from the mixed oxides. The ammonia group oxides are then fused in pyrosulphate and the analysis continued.

The main analysis is now completed with a double precipitation of magnesium as phosphate and ignition and weighing of the precipitate to the pyrophosphate.

Alkalis are determined flame photometrically after hydrofluoric, nitric and sulphuric acid treatment and filtration.

Phosphate is determined by precipitation as ammonium phosphomolybdate followed by titration with sodium hydroxide.

Many other minor ions may be determined, the most common being sulphate, chloride and carbonate; fluoride is also known to be present and may be determined by pyrohydrolysis.

REAGENTS

Additional Reagents
Ammonium acetate

Prepared Reagents

Ammonium acetate (approx. 10%): Dilute 140 ml of acetic acid (glacial) to 2000 ml with water and add carefully 140 ml of ammonia solution ($d = 0.88$). Mix, cool and adjust to approximately pH 6, either with acetic acid or ammonia solution.

Ammonium molybdate (80 g/litre): Dissolve 80 g of ammonium molybdate in 1 litre of water; filter if necessary. Store in a polythene bottle and discard after 4 weeks or earlier if any appreciable deposit of molybdic acid is observed.

Ammonium molybdate (acid solution): Dissolve 124 g of ammonium molybdate in 14 ml of ammonia solution ($d = 0.88$) and 465 ml of water. Pour the solution slowly, with vigorous stirring, into a mixture of 400 ml of nitric acid ($d = 1.42$) and 600 ml of water. Allow to cool, stand overnight and decant the clear liquid into a polythene bottle. Discard after 4 weeks or earlier if any appreciable deposit of molybdic acid is observed.

Ammonium nitrate (neutral) (50 g/litre): Dissolve 50 g of ammonium nitrate in water, neutralize to methyl red with diluted ammonia solution (1 + 1) and dilute to 1 litre.

Ammonium phosphate (100 g/litre): Dissolve 2 g of di-ammonium hydrogen orthophosphate in 20 ml of water. This solution should be freshly prepared.

Bromine water (saturated): Shake 20 ml of bromine with 500 ml of water in a glass-stoppered bottle.

1:10-phenanthroline (10 g/litre): Prepare enough solution for immediate use at a concentration of 0.1 g of 1:10-phenanthroline hydrate in 10 ml of diluted acetic acid (1 + 1).

Silver nitrate (20 g/litre): Dissolve 5 g of silver nitrate in 250 ml of water. Store in a dark bottle.

G

Silver nitrate (1 g/litre) (Used for testing for chlorides): Dissolve 0·1 g of silver nitrate in 100 ml of diluted nitric acid (1 + 9).

Stannous chloride (10 g/litre): Dissolve, by warming, 1 g of stannous chloride in 1·5 ml of hydrochloric acid ($d = 1·18$). Cool and dilute to 100 ml. The solution should not be kept for more than 24 h.

Indicators

Bromophenol blue
2: 4-Dinitrophenol
Phenolphthalein

Standard Solutions

Nitric acid (approx. 1N)
Sodium hydroxide (approx. 1N)

PREPARATION OF SAMPLE

The sample prepared for analysis should be ground to pass completely through a 120-mesh B.S. sieve. A non-metallic (e.g. nylon bolting-cloth) sieve is preferable.

Most bone ashes can be ground roughly in a porcelain mortar and the final grinding completed in an agate mortar.

NOTE: Bone ashes, when freshly calcined, tend to contain calcium oxide which will become progressively converted to the carbonate on standing. It is preferable therefore to proceed with the analysis as rapidly as possible after grinding, and to weigh off all the portions of the sample at the same time.

PROCEDURE

Determination of Loss on Ignition

The loss on ignition should not be carried out before the rest of the analysis is started, because the free lime in the bone ash tends to pick up carbon dioxide from the atmosphere thus increasing the loss on ignition with time of standing.

Weigh 1·000 g of the finely ground, dried (110°C) sample into a platinum crucible. Place the crucible in a muffle furnace and slowly raise the temperature to 1000 ± 25°C. Ignite to constant weight at this temperature: 30 min is usually sufficient.

For routine work it is possible to start the ignition over a low mushroom flame, slowly increasing the temperature to full heat over a period of about 20 min after which the crucible is transferred to a furnace at 1000°C for 30 min.

Determination of Silica

Decomposition of the Sample

Weigh 1·000 g of the finely ground, dried (110°C) sample into a 100-ml beaker and add about 25 ml of water. Cover the beaker with a clock glass and introduce 10 ml of hydrochloric acid ($d = 1 \cdot 18$) through the lip of the beaker.

Heat to boiling and boil until solution appears to be complete. Filter off the insoluble residue on a No. 40 Whatman paper and wash free from chlorides with hot water. Reserve the filtrate and washings for the determination of silica.

Transfer the paper and residue to a platinum crucible and dry the residue and char the paper. Burn off the carbon at a low temperature and fuse the residue with about 1 g of anhydrous sodium carbonate. Dissolve the melt in diluted hydrochloric acid $(1 + 1)$ and add the solution to that reserved for the determination of silica.

Determination of Silica

Transfer the solution reserved for the determination of silica to a platinum or porcelain evaporating basin, add 1 ml of diluted sulphuric acid $(1 + 1)$ and evaporate to dryness on a steam bath, breaking up from time to time the crust that forms and hinders evaporation.

When the residue is completely dry, cover the basin with a clock glass and bake in an air oven at 110°C for 1 h.

Allow to cool, and drench the residue with 10 ml of hydrochloric acid ($d = 1 \cdot 18$). After standing for a few minutes, add 50 ml of hot water and digest on a steam bath for 5 min to dissolve the salts, then filter through a No. 42 Whatman paper.

Transfer the silica to the filter with a jet of hot, diluted hydrochloric acid $(1 + 19)$ and scrub the basin with a "bobby". Wash the residue five times with hot, diluted hydrochloric acid $(1 + 19)$ and then with hot water until it is free from chlorides. Reserve the filtrate and washings for the determination of residual silica.

Place the residue and paper, without drying, in a weighed platinum crucible and heat cautiously to dry the residue and char the paper. Then burn off the carbon at a low temperature and finally ignite at 1200°C for 30 min and then to constant weight. Allow to cool, and weigh to obtain the weight of impure silica.

Moisten the weighed residue with 10 drops of diluted sulphuric acid $(1 + 1)$ and add about 5 ml of hydrofluoric acid (40% w/w). Evaporate to dryness on a sand bath in a fume cupboard. Ignite the dry residue at 1200°C for 5 min, allow the crucible to cool and weigh.

Subtract the weight of this residue from the weight of the impure silica to obtain the gravimetric silica.

Determination of Residual Silica

Transfer the filtrate and washings from the gravimetric silica to a 500-ml volumetric flask and dilute to 500 ml.

Transfer 5 ml of this solution to a 100-ml volumetric flask A, add 15 ml of water and 2 drops of 2:4-dinitrophenol indicator. Add diluted ammonia solution $(1 + 1)$ dropwise until the indicator turns yellow (note the amount of ammonia used), then add 5 ml of hydrochloric acid $(1 + 4)$.

To another 100-ml volumetric flask B, add 20 ml of water and the same amount of ammonia solution (1 + 1) as used to neutralize the aliquot in flask A. Add 2 drops of 2:4-dinitrophenol indicator followed by diluted hydrochloric acid (1 + 4) until the solution is neutral and then 5 ml in excess.

To both flasks add 6 ml of ammonium molybdate solution (80 g/litre) and stand for 5–10 min at a temperature of not less than 20°C and not greater than 30°C. Then add, with swirling, 45 ml of diluted hydrochloric acid (1 + 1) and stand for 10 min.

Add 10 ml of stannous chloride solution (10 g/litre), dilute to 100 ml and mix. Measure the optical density of the solution in flask A against the solution in flask B in 10-mm cells at 800 nm, or by using a colour filter (Ilford 609) in a suitable instrument.

The colour is stable between 5 and 30 min after the addition of the stannous chloride solution.

Determine the silica content of the solution by reference to a calibration graph.

Calculation of Total Silica Content

Add the residual silica content to the figure obtained for the "gravimetric" silica to obtain the total silica content.

Determination of Alumina, Ferric Oxide, Titania, Lime and Magnesia

Decomposition of the Sample

Weigh 1·000 g of the finely ground, dried (110°C) sample into a 100-ml beaker and add about 25 ml of water. Cover the beaker with a clock glass and introduce 10 ml of hydrochloric acid ($d = 1·18$) through the lip of the beaker. Heat to boiling and boil until solution appears to be complete.

Filter off the insoluble residue on a No. 40 Whatman paper and wash free from chlorides with hot water. Reserve the filtrate and washings for the determination of lime.

Transfer the paper and residue to a platinum crucible and dry the residue and char the paper. Burn off the carbon at a low temperature and cool.

Add 10 ml of water, 10 ml of diluted sulphuric acid (1 + 1), and 5 ml of hydrofluoric acid (40 % w/w). Evaporate the solution on a sand bath in a fume cupboard until strong fumes are evolved.

Cool the crucible, add 5 ml of diluted hydrochloric acid (1 + 1) and dissolve the residue by warming. Transfer the solution reserved for the determination of lime to a platinum basin and add the contents of the crucible.

Add 10 ml of diluted sulphuric acid (1 + 1) and about 5 ml of hydrofluoric acid (40 % w/w). Evaporate until strong fumes are evolved, cool, dilute with water and again evaporate to fumes.

Determination of Lime

Transfer the contents of the basin to a 600-ml beaker and dilute to 100 ml with water, boil for 10 min and allow to cool.

Add 350 ml of ethanol (95%), with stirring, and allow the precipitate to stand for at least 4 h or preferably overnight.

Filter through a No. 42 Whatman paper and wash four times with ethanol (95%). Reserve the filtrate (F) and washings for the determination of ferric oxide, titania and alumina.

Transfer the bulk of the precipitate back to the precipitation beaker with a jet of hot water. Transfer the paper and residue to a platinum crucible, heat gently to dry the residue and char the paper, then burn off the carbon at a low temperature.

Dissolve the residue in 2 ml of hydrochloric acid ($d = 1\cdot18$), transfer the solution to the precipitation beaker and dissolve the remainder of the precipitate in 25 ml of hydrochloric acid ($d = 1\cdot18$) and 100 ml of hot water. Boil until solution appears to be complete, dilute to about 200 ml and allow to stand overnight.

If any residue is seen, filter the solution through a No. 40 Whatman paper, wash four times with hot water, transfer the paper and residue to a platinum crucible and heat, to dry the residue and char the paper. Burn off the carbon at a low temperature and fuse the residue with 1 g of anhydrous sodium carbonate. Dissolve the melt in 5 ml of diluted hydrochloric acid $(1 + 1)$ and add the solution to the previous filtrate.

Add about 5 ml of bromine water to the solution which has stood overnight or to the combined solutions prepared as in the previous paragraph, as the case may be. Boil to expel carbon dioxide, cool slightly and add diluted ammonia solution $(1 + 1)$ until the solution is just alkaline to bromophenol blue. Filter off any traces of ferric oxide, titania or alumina which may have co-precipitated with the calcium sulphate, on a No. 40 Whatman paper and wash eight times with hot ammonium chloride solution (10 g/litre). Reserve the filtrate for the determination of lime.

Dissolve the precipitate in 10 ml of hydrochloric acid ($d = 1\cdot18$) and wash the paper thoroughly with hot water. Add the solution and washings to the filtrate (F), and discard the paper.

Adjust the volume of the solution reserved for the determination of lime to about 250 ml, acidify with hydrochloric acid ($d = 1\cdot18$), and add 2 g of oxalic acid. Boil the solution, then add with stirring, diluted ammonia solution $(1 + 1)$ until the solution is just alkaline to bromophenol blue, followed by an excess of 10 ml of diluted ammonia solution $(1 + 1)$.

Cover the beaker with a clock glass and digest on a steam bath for 2 h. Allow to cool and stand for 1 h.

Filter through a Buchner funnel with a sintered-glass mat (porosity 4) or a No. 42 Whatman paper and wash the precipitate eight times with cold ammonium oxalate solution (10 g/litre). Reserve the filtrate and washings for the extraction of magnesia.

Dissolve the precipitate through the filter with 20 ml of hot, diluted hydrochloric acid $(1 + 1)$ and wash thoroughly with hot water. Transfer the solution and washings back to the precipitation beaker.

Add about 0·5 g of oxalic acid, boil the solution and precipitate the calcium as before. Allow to cool and stand for 1 h. Filter through a No. 42 Whatman paper and wash thoroughly with cold ammonium oxalate solution (10 g/litre). Add the filtrate and washings to that reserved for the extraction of magnesia.

Place the precipitate and paper in a weighed platinum crucible and heat gently to dry the residue and char the paper. Burn off the carbon at a low temperature and finally ignite at 1000°C for 30 min and then to constant weight. Allow to cool over a good desiccant, because the precipitate is very hygroscopic, and keep the crucible covered with a tight-fitting lid. Weigh rapidly to obtain the weight of the calcium oxide.

Extraction of Magnesia

As it is possible that a little magnesia has been precipitated with the calcium sulphate this must be separated from the filtrate from the determination of lime.

Acidify the combined solution reserved for the extraction of magnesia with hydrochloric acid ($d = 1 \cdot 18$) and evaporate to about 300 ml. Add 20 ml of freshly prepared ammonium phosphate solution (100 g/litre).

Heat the solution to boiling and make just alkaline to bromophenol blue with diluted ammonia solution ($1 + 1$). Cool to room temperature and add 20 ml of ammonia solution ($d = 0 \cdot 88$). Stir vigorously to start the precipitation and allow to stand overnight at a low temperature.

Filter through a No. 42 Whatman paper and wash eight times with cold, diluted ammonia solution ($1 + 39$). Discard the filtrate and washings. Dissolve the precipitate through the filter with 20 ml of hot, diluted hydrochloric acid ($1 + 1$) and wash thoroughly with hot water. Add the solution and washings to the filtrate (F).

Determination of Ferric Oxide, Titania and Alumina

Evaporate the combined filtrate (F) to a volume of about 150 ml on a steam bath or hot plate (avoid a naked flame until all the alcohol has been removed) and transfer it to a platinum basin. Evaporate on a sand bath until strong fumes of sulphur trioxide are evolved. Cool, then add 10 ml of nitric acid ($d = 1 \cdot 42$) and evaporate until most of the sulphuric acid has been removed.

Cool, dilute cautiously with water to about 100 ml and add 10 ml of hydrochloric acid ($d = 1 \cdot 18$). Transfer the solution to a 400-ml beaker, boil for 10 min, filter through a No. 40 Whatman paper and wash thoroughly with hot water. Discard the paper and residue.

Dilute the filtrate and washings to about 300 ml, add 5 ml of bromine water, (saturated), to oxidize the iron, and boil for a few minutes.

Add 10 ml of freshly prepared ammonium phosphate solution (100 g/litre), followed by diluted ammonia solution ($1 + 1$) until the solution is faintly alkaline to bromophenol blue, make just acid with hydrochloric acid ($d = 1 \cdot 18$) and add 1 ml of diluted hydrochloric acid ($1 + 1$) in excess.

Heat the solution to boiling, add 30 ml of ammonium acetate solution (250 g/litre) and boil for 5 min. Filter through a No. 540 Whatman paper and transfer the precipitate to the filter with hot, neutral ammonium nitrate solution (50 g/litre), scrubbing the beaker with a "bobby". Wash the precipitate free from chlorides with hot, neutral ammonium nitrate solution (50 g/litre).

Reserve the filtrate and washings for the determination of magnesia. If the precipitate is large (greater than about 1%) the precipitation should be repeated.

Place the paper and residue in a weighed platinum crucible and heat gently to dry the residue and char the paper. Burn off the carbon at a low temperature and finally ignite at 1000°C for 20 min. Cool and weigh as the mixed phosphates.

(Platinum crucibles may be seriously attacked during the ignition if the precipitate is not thoroughly washed with ammonium nitrate solution, or if the precipitate is ignited too strongly before all the carbon is removed.)

Because the presence of phosphoric acid tends to bleach the colour produced with 1:10-phenanthroline, there will be a tendency to obtain low results for iron. For normal bone ashes, where the iron content and the total R_2O_3 content are low, this effect can be ignored unless results of the highest accuracy are required. If the iron content is high or where the total ammonia group oxides exceed about 1%, it is usually advisable to grind the weighed mixed phosphates with anhydrous sodium carbonate in an agate mortar, and fuse in the same crucible and extract with hot water. Filter off the residual iron and titanium, and wash thoroughly with hot sodium carbonate solution (10 g/litre) discarding the filtrate. Transfer the paper and precipitate back to the platinum crucible, burn off the carbon and continue the analysis as described below by fusing with potassium pyrosulphate etc.

Fuse the ignited, weighed phosphates in the same platinum crucible with 10 g of potassium pyrosulphate. Cool the crucible and extract the melt in a beaker containing about 150 ml of water to which 10 ml of sulphuric acid ($d = 1.84$) has been cautiously added.

Cool and dilute the solution to 250 ml in a volumetric flask. The solution thus obtained will be referred to as the "stock test solution".

Determination of ferric oxide

Transfer a 5-ml aliquot of the stock test solution to a 100-ml volumetric flask. Add 2 ml of hydroxyammonium chloride solution (100 g/litre), 5 ml of 1:10-phenanthroline solution (10 g/litre) and 2 ml of ammonium acetate solution (approx. 10%). Allow to stand for 15 min, dilute to 100 ml and mix.

Measure the optical density of the solution against water in 10-mm cells at 510 nm or by using a colour filter (Ilford 603) in a suitable instrument. The colour is stable between 15 min and 75 min after the addition of the ammonium acetate solution.

Determine the ferric oxide content of the solution by reference to a calibration graph.

Determination of titania

Transfer a 40-ml aliquot of the stock test solution to each of two 100-ml volumetric flasks, A and B. To each flask, add 10 ml of diluted phosphoric acid (2 + 3) and, to flask A only, 10 ml of hydrogen peroxide solution (6%). Dilute the solution in each flask to 100 ml and shake well.

Measure A against B in 40-mm cells at 398 nm or by using a colour filter (Ilford 601) in a suitable instrument. The colour is stable between 5 min and 24 h after the addition of the hydrogen peroxide solution.

Determine the titania content of the solution by reference to a calibration graph.

Calculation of alumina

The alumina content of the material is calculated from the weight of the mixed phosphates and the ferric oxide and titania contents.

Calculate the equivalent weight of the ferric phosphate from the ferric oxide and of titanium phosphate from the titania as follows:

$$\text{Weight of } Fe_2O_3 \times 1.8889 = \text{weight of } FePO_4$$
$$\text{Weight of } TiO_2 \times 2.7766 = \text{weight of } TiP_2O_7.$$

Add the weights of the phosphates of iron and titanium together and subtract from the total weight of mixed phosphates to obtain the weight of the aluminium phosphate.

$$\text{Weight of } AlPO_4 \times 0.4180 = \text{weight of } Al_2O_3.$$

Determination of Magnesia

Acidify the solution from the determination of ferric oxide, titania and alumina with hydrochloric acid ($d = 1.18$) and evaporate to about 250 ml. Heat to boiling, add diluted ammonia solution $(1 + 1)$, with stirring, until the solution is alkaline to bromophenol blue and cool to room temperature. Add 20 ml of ammonia solution ($d = 0.88$), stir vigorously to start the precipitation and allow to stand overnight at a low temperature.

Filter through a No. 42 Whatman paper and wash four times with cold, diluted ammonia solution $(1 + 39)$, discarding the filtrate and washings.

Dissolve the precipitate through the filter with 20 ml of hot, diluted nitric acid $(1 + 1)$ and wash thoroughly with hot water. Transfer the filtrate and washings back to the precipitation beaker.

Add 1 ml of ammonium phosphate solution (100 g/litre), heat the solution to boiling and, while stirring, make just alkaline to bromophenol blue with diluted ammonia solution $(1 + 1)$.

Cool, add 10 ml of ammonia solution ($d = 0.88$) and stir vigorously to start the precipitation. Allow to stand overnight at a low temperature.

Filter through a No. 42 Whatman paper, transfer the precipitate to the filter with a jet of cold, diluted ammonia solution $(1 + 39)$, scrubbing the beaker with a "bobby". Wash thoroughly with cold, diluted ammonia solution $(1 + 39)$, discarding the filtrate and washings.

Transfer the precipitate and paper to a weighed platinum crucible and heat gently to dry the precipitate and char the paper. Burn off the carbon at a low temperature. (Platinum crucibles may be seriously attacked during the ignition if the precipitate is not thoroughly washed with diluted ammonia solution, or if the precipitate is ignited too strongly before all the carbon is oxidized.) Finally ignite at 1150°C for 15 min, cool and weigh the residue as magnesium pyrophosphate.

Calculation—Weight of $Mg_2P_2O_7 \times 0.3623$ = weight of MgO.

Determination of Alkalis

Determine the alkalis in the sample flame-photometrically by the method described in Chapter 16 under "Bone Ash"

Determination of Phosphate

Weigh 0.2000 g of the finely ground, dried (110°C) sample into a 100-ml beaker. Cover the beaker with a clock glass and introduce 20 ml of diluted nitric acid (1 + 3) through the lip of the beaker.

Heat to boiling and boil until solution appears to be complete. Filter off the insoluble residue on a No. 40 Whatman paper and wash thoroughly with hot water. Reserve the filtrate and washings for the determination of phosphate.

Transfer the paper and residue to a platinum crucible, heat gently to dry the residue and char the paper. Burn off the carbon at a low temperature. Add 10 ml of nitric acid (d = 1.42) and about 5 ml of hydrofluoric acid (40% w/w) and evaporate to dryness. Add a further 10 ml of nitric acid (d = 1.42) and again evaporate to dryness. Add 5 ml of nitric acid (d = 1.42) and evaporate to dryness for a third time.

Treat the residue with 10 ml of diluted nitric acid (1 + 1) and evaporate to about 5 ml. Add the solution to the filtrate and washings reserved for the determination of phosphate, cover the beaker with a clock glass and boil the solution for 15 min.

If necessary, filter through a No. 540 Whatman paper and wash with hot water until free from acid. Discard the residue.

Transfer the solution, or filtrate and washings, to a 500-ml conical flask and evaporate down to about 50 ml. Cool and add ammonia solution (d = 0.88) until a slight permanent precipitate is formed. Add nitric acid (d = 1.42), drop by drop, until the precipitate is just redissolved and then add a further 4 ml of nitric acid (d = 1.42).

Heat the solution to about 75°C remove from the source of heat and add 100 ml of ammonium molybdate (acid solution), previously heated to 55°C, in a slow, steady stream with vigorous shaking. Insert a rubber bung into the neck of the flask and shake the flask and contents for 2 min. Maintain at a temperature of 50°C for 30 min.

Allow the flask to cool and stand for at least 3 h, but not longer than overnight.

Filter the solution through a crucible with a sintered-glass mat (porosity 4) or a No. 42 Whatman paper. Transfer the precipitate to the filter, and wash with neutral potassium nitrate solution (1 g/litre). Continue the washing until 10 ml of the washings give a strong alkaline reaction with 1 drop of standard sodium hydroxide solution (approx. 1N) and 1 drop of phenolphthalein indicator.

G*

Transfer the paper, or crucible and precipitate, to a 500-ml flask or beaker respectively and introduce 30 ml (or more, if necessary) of standard sodium hydroxide solution (approx. 1N). Shake the flask or beaker and contents until the precipitate has dissolved.

Add a few drops of phenolphthalein indicator and titrate the residual sodium hydroxide with standard nitric acid (approx. 1N).

Calculation—Correct the titration figures to exactly normal.

Weight of P_2O_5 = 0·003092 × (ml of 1·0N NaOH − ml of 1·0N HNO_3 used in back titration).

Determination of Sulphate

Weigh 1·000 g of the finely ground, dried (110°C) sample into a 250-ml beaker. Add 75 ml of water and 10 ml of bromine water. Cover with a clock glass and add 10 ml of hydrochloric acid (d = 1·18) through the lip of the beaker. Heat to boiling, and boil until solution appears to be complete. Filter through a No. 40 Whatman paper and wash thoroughly with hot water. Reserve the filtrate and washings for the determination of sulphate.

Transfer the residue and paper to a platinum crucible and heat gently to dry the residue and char the paper. Burn off the carbon at a low temperature.

Fuse the residue in about 1 g of anhydrous sodium carbonate. Allow to cool and dissolve the melt in diluted hydrochloric acid (1 + 1).

Add the solution to the filtrate and washings reserved for the determination of sulphate and boil to expel carbon dioxide.

While boiling, add 5 ml of barium chloride solution (100 g/litre), drop by drop, continue boiling for 2 min, transfer to a steam bath for 1 h and allow to cool. Allow to stand overnight.

Filter through a No. 42 Whatman paper, transfer the precipitate to the paper with a jet of hot water and scrub the beaker with a "bobby". Wash thoroughly with hot water until free from chlorides.

Transfer the precipitate and paper to a weighed platinum crucible and heat gently to dry the residue and char the paper. Burn off the carbon at a low temperature.

Cool and moisten the residue with a few drops of diluted sulphuric acid (1 + 1), add about 5 ml of hydrofluoric acid (40 % w/w) and evaporate slowly to dryness on a sand bath in a fume cupboard.

Ignite the dry residue at 1000°C for 30 min. Allow to cool and weigh.

Calculation—Weight of $BaSO_4$ × 0·3430 = weight of SO_3.

Determination of Chloride

Weigh 2·000 g of the finely ground, dried (110°C) sample into a 250-ml beaker, and add 50 ml of water. Cover the beaker with a clock glass and introduce 10 ml of nitric acid (d = 1·42). Heat to boiling and boil until solution appears to be complete.

Filter off the residue on a No. 40 Whatman paper and wash thoroughly with hot water. Discard the residue and paper, boil the filtrate to remove carbon dioxide and allow to cool.

The subsequent precipitation should be carried out in a subdued light, as silver chloride is sensitive to light.

Add, drop by drop, 5 ml of silver nitrate solution (20 g/litre) and heat the solution to boiling and boil for 5 min. Make certain that precipitation is complete by adding, if necessary, a further few drops of silver nitrate solution. Stand the solution overnight in the dark.

Filter through a weighed crucible with a sintered-glass mat (porosity 4), transferring the precipitate to the filter with a jet of cold, diluted nitric acid (1 + 49) and scrubbing the beaker with a "bobby". Wash thoroughly with cold, diluted nitric acid (1 + 49). Dry the crucible and contents in an air oven at 110°C for 2 h, cool and weigh.

Calculation—Weight of AgCl × 0·2474 = weight of Cl.

Determination of Carbonate

The method to be used for the determination of carbonate is described in Chapter 19.

CHAPTER 15

ANALYSIS OF FRITS AND GLAZES
(Containing PbO and/or B_2O_3)

GENERAL

Frits and glazes are conveniently grouped together as one class of materials, but they do, in fact, vary appreciably in composition. They may contain only oxides that are present in aluminosilicates with the addition of borate as in borax frits or with the addition of lead as in lead bisilicates. On the other hand they may be materials of great complexity containing opacifying and colouring oxides.

This chapter is concerned only with the analysis of colourless or white opacified frits and glazes. The analysis of the extremely wide variety of colours is beyond simple description; such materials can only be analysed individually, taking into account the results of a preliminary qualitative or spectrographic analysis and devising an appropriate method.

Borax frits which contain only those elements normally found in alumino-silicates together with boron, can best be analysed by the coagulation method (Chapter 5), bearing in mind that borate is already present in the sample and does not therefore need to be added to the flux. Lead bisilicate frits also utilize a specific method, but for other glazes a basically classical method is described.

The chapter thus contains descriptions of the analysis of:

A. Borax frit—details of full method, including the determination of boric oxide and the zinc uranyl acetate method for soda.
B. Lead bisilicate frit—details of full method.
C. Glazes containing elements such as Pb, Zn, Sn, Ba, Zr by a basically classical method.
D. Glazes as above, giving suggested methods in outline for various combinations of elements by more modern methods.

A. Method for the Analysis of Borax Frits

PRINCIPLE OF THE METHOD

Loss on Ignition

As the material melts below 1000°C it is customary to carry out the loss on ignition determination at a lower temperature, viz. 800°C.

Determination of Silica, Titania and Alumina

The sample is fused with fusion mixture and dissolved in hydrochloric acid and a small amount of sulphuric acid added. The mixture is transferred to a steam bath until a stiff gel is formed, coagulated with a polyethylene oxide resin and the silica separated by filtration. The crucible containing this silica is ignited and weighed before and after treatment with hydrofluoric and sulphuric acids. The silica remaining in the filtrate is subsequently determined by a spectrophotometric method, based on the formation of silicomolybdic acid with or without reduction to molybdenum blue, in an aliquot of the solution used for the determination of titania and alumina.

The use of a small ratio of flux to sample enables the fusion, acid decomposition and evaporation to be carried out in one and the same platinum dish.

The residue from the silica is fused with potassium pyrosulphate and the solution of the melt added to the filtrate from the silica. After dilution to a standard volume, the titania is determined spectrophotometrically with hydrogen peroxide and the alumina volumetrically with diaminoethanetetraacetic acid disodium salt dihydrate (EDTA) and zinc, after removal of ferric and titanium ions by a cupferron/chloroform solvent extraction. Sulphuric acid is added to the solution before the extraction to counteract interference with the end-point by the high lime: alumina ratio of the material.

Determination of Ferric Oxide, Lime, Magnesia, Potash and Soda (Routine Method)

The sample is decomposed with hydrofluoric, sulphuric and nitric acids to remove the silica. A further evaporation with sulphuric and nitric acids removes traces of fluoride, the residue is dissolved in nitric acid and the solution diluted to a standard volume.

Ferric oxide is determined spectrophotometrically with 1:10-phenanthroline hydrate. Lime and magnesia are determined volumetrically with EDTA, measured volumes being withdrawn and interfering elements being complexed by the addition of triethanolamine. The solutions are made strongly alkaline and then titrated for lime, and lime plus magnesia; Calcein and methylthymol blue complexone, respectively, being used as indicators. Alkalis are determined flame photometrically.

Determination of Soda (Accurate Method)

The sample is decomposed with hydrofluoric and perchloric acids. After evaporation the residue is dissolved in hydrochloric acid and the sodium in the whole solution precipitated and weighed as sodium zinc uranyl acetate.

Determination of Boric Oxide

The sample is decomposed by fusion with fusion mixture, followed by aqueous decomposition of the melt. After acidification with hydrochloric acid, the solution is neutralized with calcium carbonate and boiled under a reflux condenser. The hot solution is filtered and, after making slightly acid, is again refluxed. The cooled solution is titrated to pH 7 with sodium hydroxide, to neutralize the free mineral acid. Mannitol is added to convert the weak boric acid to the mannitol-boric acid complex which is then titrated back to pH 7.

REAGENTS

Additional Reagents

Mannitol
Uranyl acetate
Zinc acetate.

Prepared Reagents

Ammonium acetate (approx. 10%): Dilute 140 ml of acetic acid (glacial) to 2000 ml with water and add carefully with stirring 140 ml of ammonia solution ($d = 0\cdot88$). Mix, cool and adjust to pH $6\cdot0$–$6\cdot5$, either with acetic acid or ammonia solution.

Ammonium acetate buffer: Add with stirring 120 ml of acetic acid (glacial) to 500 ml of water, followed by 74 ml of ammonia solution ($d = 0\cdot88$). Cool, dilute to 1 litre and mix.

Ammonium molybdate (80 g/litre): Dissolve 80 g of ammonium molybdate in water, filter if necessary, dilute to 1000 ml and mix. Store in a polythene bottle. Discard after 4 weeks, or earlier if any appreciable deposit is observed.

Cupferron (60 g/litre): Dissolve $1\cdot5$ g of cupferron in 25 ml of water and filter. This solution must be freshly prepared.

If the reagent is discoloured or gives a strongly coloured solution a new stock should be obtained. The solid reagent should be stored in a tightly stoppered bottle in the presence of a piece of ammonium carbonate to prevent decomposition.

1:10-Phenanthroline hydrate (10 g/litre): Prepare enough solution for immediate use at a concentration of $0\cdot1$ g of 1:10-phenanthroline hydrate in 10 ml of acetic acid $(1 + 1)$.

Polyethylene oxide (2·5 g/litre): Add $0\cdot5$ g of polyethylene oxide* to 200 ml of water slowly, with stirring, preferably on a mechanical stirrer, until dissolved. Discard after 2 weeks.

Sodium zinc uranyl acetate wash solution: Transfer 475 ml of ethanol (95%) and 25 ml of water to a flask of about 1 litre capacity. Add 1 g of sodium zinc uranyl acetate, then heat the flask and contents carefully to about 60°C and shake to mix them thoroughly. Cool the solution and allow it to stand for at least 24 h. When required for use, ensure that the solution is at room temperature and filter through a sintered-glass crucible of porosity grade No. 4 or a No. 42 Whatman filter paper immediately before use.

Stannous chloride (10 g/litre): Dissolve, by warming, 1 g of stannous chloride in $1\cdot5$ ml of hydrochloric acid ($d = 1\cdot18$). Cool and dilute to 100 ml. This solution should not be kept longer than 24 hours.

Sulphuric–nitric acid mixture: To 650 ml of water, add 100 ml of diluted sulphuric acid $(1 + 1)$ and 250 ml of nitric acid ($d = 1\cdot42$).

*Union Carbide Polyox Resins WSR 35, WSR N-80, WSR 205, WSR N-750 or WSR N-3000 are suitable as sources of polyethylene oxide.

Zinc uranyl acetate: Transfer 100 g of uranyl acetate and 500 g of zinc acetate to a flask of about 2 litres capacity. Dilute 27 ml of acetic acid (glacial) to 100 ml with water, pour the diluted acid into the flask and add a further 800 ml of water. Shake the flask to ensure thorough mixing and heat the mixture just to boiling to effect solution of the salts. Add 0·1 g of sodium chloride, shake the flask well to dissolve the sodium chloride, set aside to cool, and allow to stand for at least 24 h. When required for use ensure that the solution is at room temperature and filter through a sintered-glass crucible of porosity grade No. 4 or a No. 42 Whatman filter paper immediately before use.

Indicators

Bromophenol blue
Calcein (screened)
2:4-Dinitrophenol
Dithizone
Methylthymol blue

Standard Solutions

EDTA 5 g/litre
EDTA 0·05M
Sodium hydroxide (approx. 0·1N)—Standardize against a sample of known borate content, making any necessary allowance for possible change in loss on ignition. The material should be taken through the full procedure (a suitable standard, Borosilicate Glass No. 93, can be obtained from the Department of Commerce, Bureau of Standards, Washington, D.C.).
Zinc 0·05M

BLANK DETERMINATIONS

A blank determination should be carried out on all reagents in accordance with the general scheme of analysis. The fusion mixture should be weighed into the platinum dish and then dissolved by the addition of hydrochloric and sulphuric acids, the fusion being omitted to avoid undue attack on the platinum dish, which occurs in the absence of the sample.

PREPARATION OF SAMPLE

The sample prepared for analysis should be ground to pass completely through a sieve equivalent to a 120 B.S. mesh. A non-metallic (e.g. nylon bolting-cloth) sieve is preferable.

PROCEDURE

Determination of Loss on Ignition

Weigh 1·000 g of the prepared, dried (110°C) sample into a porcelain crucible. Raise the temperature slowly and finally ignite at 800°C for 30 min, with free access to air to maintain oxidizing conditions, cool and weigh.

Determination of Silica, Titania and Alumina

Decomposition of the Sample

Weigh 1·000 g of the prepared, dried (110°C) sample into a platinum dish; a convenient size is 70-mm dia., 40-mm deep, of effective capacity 75 ml. Add 3 g of fusion mixture and mix intimately to form a charge of about 50-mm dia. in the centre of the dish. Cover the dish with a lid.

Heat over a gas burner, cautiously at first until frothing ceases. Then heat over a Méker burner until a clear melt is obtained. Remove the dish from the source of heat and allow to cool.

Add 15 ml of hydrochloric acid ($d = 1·18$), 1 ml of diluted sulphuric acid (1 + 1), 10 ml of water and transfer the dish to a steam bath to facilitate decomposition of the melt. If the solution gels before dissolution is complete it may be necessary to stir the gel gently with a glass rod. When solution is complete, remove the lid and wash any spray adhering to it into the dish with the minimum amount of water.

Determination of the Main Silica

Allow the dish to remain on the steam bath until a stiff gel is formed, thoroughly mix a Whatman accelerator tablet into the gel until the mixture is homogeneous, ensuring that the gel adhering to the sides of the dish is also broken up.

Add slowly, with stirring, 5 ml of polyethylene oxide coagulant solution (2·5 g/litre) and mix well, ensuring that any gel adhering to the sides of the dish is brought into contact with the coagulant. Add 10 ml of water, mix and allow to stand for 5 min.

Filter through a 125-mm No. 42 Whatman paper, transferring the silica to the filter with hot hydrochloric acid (1 + 19), scrubbing the dish with a rubber-tipped glass rod. Wash the precipitate five times with hot hydrochloric acid (1 + 19) and then with hot water until free from chlorides. Reserve the filtrate and washings.

Transfer the paper and precipitate to an ignited and weighed platinum crucible. Ignite at a low temperature until the precipitate is free from carbonaceous matter and then heat in a muffle furnace at 1200°C to constant weight, 30 min being normally sufficient. The ignition should be started at a lower temperature than usual or a noticeable amount of carbon may remain after ignition. In some cases a trace of carbon may remain but its weight appears to be negligible and does not affect the results.

Moisten the contents of the cold crucible with water, add 5 drops of sulphuric acid (1 + 1) and 10 ml of hydrofluoric acid (40% w/w). Evaporate to dryness on a sand bath in a fume cupboard. For the evaporation, the crucible and contents should be heated from below; the use of top heating alone, as with a radiant heater, can result in incomplete elimination of silica by the hydrofluoric acid.

Heat the crucible and residue, cautiously at first, over a gas flame and finally for 5 min at 1200°C, cool and weigh. The difference between the two weights represent the "gravimetric" silica.

Preparation of the Solution for the Determination of Residual Silica, Titania and Alumina

Fuse the residue from the hydrofluoric acid treatment of the "gravimetric" silica with 1 g of potassium pyrosulphate, dissolve the melt in water containing a few drops of hydrochloric acid ($d = 1\cdot18$) and add the solution to the reserved filtrate. Cool, dilute the combined solution to 500 ml in a volumetric flask and mix.

This solution is referred to as the "stock" solution.

Determination of Residual Silica

The presence of phosphates, vanadates, etc., which produce a yellow colour with ammonium molybdate, will result in positive errors in the silica determination. Borax frits rarely contain sufficient of these constituents to justify account being taken of them in the analysis procedure. The molybdenum blue method eliminates interference from phosphate.

Silicomolybdate method

Transfer a 20-ml aliquot of the "stock" solution to each of two 50-ml volumetric flasks, A and B, and add a few drops of 2:4-dinitrophenol indicator. To flask B add ammonia solution ($d = 0\cdot88$) dropwise until the indicator turns yellow (note the number of drops used) then add 5 ml of hydrochloric acid (1 + 4). To flask A add 5 ml of hydrochloric acid (1 + 4) and the same number of drops of ammonia as needed to neutralize the solution in flask B, then add 5 ml of ammonium molybdate solution (80 g/litre).

Dilute the solution in each flask to 50 ml and shake well. Measure the silicomolybdate colour in flask A against the solution in flask B in 40-mm cells at 440 nm or by using a colour filter (Ilford 601) in a suitable instrument.

Measure the optical density not sooner than 5 min and not later than 15 min after the addition of the ammonium molybdate solution.

Determine the silica content of the solution by reference to a calibration graph.

If a cloudy solution develops in carrying out the above method, the molybdenum-blue procedure should be used.

Molybdenum-blue method

Transfer $5\cdot0$ ml of the stock solution to a 100-ml volumetric flask A, add 15 ml of water and 2 drops of 2:4-dinitrophenol indicator solution. Add diluted ammonia solution (1 + 1) dropwise until the indicator turns yellow noting the amount of ammonia used, then add 5 ml of diluted hydrochloric acid (1 + 4).

To another 100-ml volumetric flask B, add 20 ml of water and the same amount of diluted ammonia solution (1 + 1) as used to neutralize the aliquot in flask A. Add 2 drops of 2:4-dinitrophenol indicator followed by diluted hydrochloric acid (1 + 4) until the solution is neutral and then 5 ml in excess.

To both flasks add 6 ml of ammonium molybdate (80 g/litre) and stand for 5–10 min at a temperature of not less than 20°C and not greater than 30°C.

Then add, with swirling, 45 ml of diluted hydrochloric acid (1 + 1) and stand for 10 min.

Add 10 ml of stannous chloride (10 g/litre), dilute to 100 ml and mix. (The appearance of a brown colour in the solution is normal and does not interfere with the determination.) Measure the optical density of the solution in flask A against the solution in flask B in 10-mm cells at 800 nm or by using a colour filter (Ilford 609) in a suitable instrument.

Calculation of the Total Silica Content

Add the residual silica content to the figure obtained for the "gravimetric" silica to obtain the total silica content.

Determination of Titania

Transfer 20·0 ml of the stock solution to each of two 50-ml volumetric flasks, A and B. To each flask add 10 ml of phosphoric acid (2 + 3) and to flask A only, 10 ml of hydrogen peroxide solution (6%).

Dilute the solution in each flask to 50 ml and shake well. Measure the optical density of the solution in flask A against that in flask B in 40-mm cells at 398 nm, or by using a colour filter (Ilford 601) in a suitable instrument. The colour is stable between 5 min and 24 h after the addition of the hydrogen peroxide solution.

Determine the titania content of the solution by reference to a calibration graph.

Determination of Alumina

Transfer 100·0 ml of the stock solution to a 500-ml separating funnel (a Squibb's type is recommended) and add 10 ml of hydrochloric acid ($d = 1·18$) and 25 ml of sulphuric acid (1 + 1).

Add 25 ml of chloroform and 5 ml of cupferron solution (60 g/litre). Stopper the funnel and shake vigorously. Release the pressure in the funnel by carefully removing the stopper and rinse the stopper and neck of the funnel with water. Allow the layers to separate and withdraw the chloroform layer. Confirm that extraction is complete by checking that the addition of a few drops of cupferron solution (60 g/litre) does not produce a permanent coloured precipitate. Add further 10-ml portions of chloroform and repeat the extraction until the chloroform layer is colourless, then wash the stem of the funnel, inside and out, with chloroform, using a polythene wash bottle.

Discard the chloroform extracts and transfer the aqueous solution to a 500-ml conical flask. Add a few drops of bromophenol blue solution and then add ammonia solution ($d = 0·88$) until the solution is just alkaline. Re-acidify quickly with hydrochloric acid ($d = 1·18$) and add 5–6 drops in excess. Add sufficient EDTA solution (0·05M) to provide an excess of a few millilitres over the expected amount; 1 ml of EDTA solution (0·05M) $\equiv 1·275\%$ Al_2O_3. Then add ammonium acetate buffer solution until the indicator turns blue followed by 10 ml in excess. Boil the solution for 10 min and cool rapidly.

Add a volume of ethanol (95%) equal to the total volume of the solution, followed by 1–2 ml of dithizone indicator and titrate with the standard zinc solution (0·05M), using a semi-micro or similar burette, from blue-green to the first appearance of a permanent pink colour.

Calculation—If the EDTA solution is not exactly 0·05M, calculate the equivalent volume of exactly 0·05M EDTA.

If V ml is the volume of EDTA (0·05M) and v ml is the volume of zinc solution (0·05M) used in the back-titration, then $Al_2O_3\% = 1·275 \, (V-v)$.

Determination of Ferric Oxide, Lime, Magnesia, Potash and Soda (Routine Method)

Decomposition of the Sample

Weigh 0·500 g of the prepared, dried (110°C) sample into a small platinum dish. Add 10 ml of sulphuric–nitric acid mixture and about 15 ml of hydrofluoric acid (40% w/w).

Cover the dish with a lid and digest the mixture in a fume cupboard on a hot sand bath to facilitate attack of the sample before evaporation; then evaporate to dryness, being careful to avoid spurting. Cool, add 10 ml of sulphuric–nitric acid mixture and rinse down the inside of the dish with water. Again evaporate carefully to dryness.

To the cool, dry residue add 20 ml of dilute nitric acid (1 + 9) and digest the contents of the dish on a steam bath until dissolved. Cool and dilute the solution to 500 ml in a volumetric flask. This is referred to as the "sample solution A".

NOTE: As sulphuric acid is used for the decomposition it may be preferable to determine lime and magnesia on the main stock solution.

Determination of Ferric Oxide

Transfer 50·0 ml of the sample solution A to each of the two 100-ml volumetric flasks, A and B. To each flask add 2 ml of hydroxyammonium chloride solution (100 g/litre) and to flask A only, 5 ml of 1:10-phenanthroline solution (10 g/litre) and ammonium acetate solution (approx. 10%) until a pink colour develops, followed by 2 ml in excess. Add the same volume of ammonium acetate solution (approx. 10%) to flask B.

Allow to stand for 15 min, dilute the solution in each flask to 100 ml and shake well. Measure the optical density of the solution in flask A against that in flask B in 40-mm cells at 510 nm, or by using a colour filter (Ilford 603) in a suitable instrument. The colour is stable between 15 and 75 min after the addition of the ammonium acetate solution.

Determine the ferric oxide content of the solution by reference to a calibration graph.

Determination of Lime

Transfer 100·0 ml of the sample solution A to a 500-ml conical flask. Add 5 ml of triethanolamine (1 + 1) and 10 ml of potassium hydroxide solution (250 g/litre) and dilute to about 200 ml. Add about 0·015 g of screened

Calcein indicator and titrate with standard EDTA solution (5 g/litre) from a 10-ml semi-micro or similar burette, the colour change being from fluorescent green to pink.

Determination of the Sum of Lime and Magnesia

Transfer 100·0 ml of the sample solution A to a 500-ml conical flask. Add 10 drops of hydrochloric acid ($d = 1·18$), 20 ml of triethanolamine ($1 + 1$) and 25 ml of ammonia solution ($d = 0·88$) and dilute to about 200 ml. Add about 0·04 g of methylthymol blue complexone indicator and titrate with standard EDTA solution (5 g/litre) from a 10-ml semi-micro or similar burette, the colour change being from blue to colourless.

Calculation of Magnesia

Subtract the volume of EDTA solution used for the titration of lime from the volume of EDTA solution used for the titration of the sum of lime and magnesia. The remainder represents the volume of EDTA solution (5 g/litre) required for the titration of the magnesia.

Determination of Potash and Soda (Routine Method)

Determine the alkalis flame-photometrically by the procedure given in Chapter 16 under "Borax Frit".

Determination of Soda (Accurate Method)

Weigh 0·100 g of the prepared, dried (110°C) sample into a small platinum dish. Moisten with water, add 5 ml of hydrofluoric acid (40 % w/w) and either gently swirl the contents of the dish, or stir with a platinum wire to assist decomposition. Add 10 drops of perchloric acid ($d = 1·54$) and evaporate gently until a syrupy mass remains. Increase the heat and continue the evaporation until fumes of perchloric acid cease and the mass is dry. Remove the dish from the source of heat as soon as the contents are dry, otherwise the perchlorates may be decomposed and solution of the residue may be difficult.

Dissolve the residue in 1 ml of hydrochloric acid ($1 + 4$), warming if necessary, then cool. Add 20 ml of zinc uranyl acetate solution and thoroughly mix the contents by gently stirring for 1–2 min. Allow the dish and contents to stand at room temperature (about 20°C) for 45 min and keep the zinc uranyl acetate stock solution and the alcoholic sodium zinc uranyl acetate wash solution at the same temperature.

Wash a sintered-glass crucible, porosity grade No. 4, with water, then three times with ethanol and finally twice with 10-ml portions of diethyl ether. Maintain suction throughout and continue until no smell of ether can be detected. Carefully wipe the crucible with a dry, fluffless cloth and leave in the balance-case for 30 min, then weigh.

Meanwhile, transfer 20 ml of zinc uranyl acetate solution to a small wash bottle and about 20 ml of the wash solution to a similar bottle. Filter off the precipitated sodium zinc uranyl acetate from the sample solution, transferring completely to the weighed, sintered-glass crucible, using small volumes of the reagent solution. Allow the crucible to drain, wash it four times with the alcoholic wash solution and then twice with 10-ml portions of diethyl ether. Maintain suction throughout and continue until no smell of ether remains. Wipe the crucible with a dry, fluffless cloth, leave in the balance-case for 30 min, then weigh. Leave the crucible in the balance-case and reweigh at intervals until constant weight is attained.

Calculation—From the corrected weight of sodium zinc uranyl acetate calculate the percentage of Na_2O in the sample.

Weight of precipitate $\times$ 0·02015 = weight of Na_2O.

Determination of Boric Oxide

Weigh 0·500 g of the prepared, dried (110°C) sample into a platinum crucible and mix intimately with 10 g of fusion mixture. Cover the crucible with a lid.

Carry out the fusion over a Bunsen burner at as low a temperature and for as short a time as is necessary to obtain a satisfactory fusion. Cool the fused cake and transfer it to a 500-ml conical flask (with a ground glass neck), washing the crucible with a minimum of hot water. Digest on a hot plate until disintegration of the cake is complete.

Cool, then add 20 ml of hydrochloric acid ($d = 1·18$) and 3 drops of nitric acid ($d = 1·42$); if the hydrochloric acid is measured into the platinum crucible it will assist in cleaning it.

Neutralize the solution by the slow addition of 6 g of calcium carbonate and then boil gently under a reflux condenser for 10 min. Filter the hot solution through a No. 41 Whatman filter paper into a 500-ml conical flask (with a ground glass neck). Wash with hot water, keeping the volume of washings as low as possible.

Make the filtrate just acid to methyl red indicator with hydrochloric acid (1 + 1); 1 drop of acid is usually sufficient. Reflux for 30 min. Cool quickly, with the reflux condenser still in position, wash the condenser sparingly and transfer the solution to an 800-ml "squat-type" beaker. Care must be taken to exclude CO_2, i.e. freshly-boiled distilled water should be used, otherwise high results will be obtained. The volume of solution should be kept as low as possible throughout the analysis.

Titration—The titration is carried out with sodium hydroxide solution (0·1N), using a pH meter in conjunction with a mechanical stirrer. First titrate the solution to pH 7 to neutralize the free mineral acid (HCl). Take the burette reading (v ml), add 20 g of mannitol to the solution and titrate back to pH 7. Again note the burette reading (V ml).

Calculation—With exactly 0·1N NaOH solution, the percentage of B_2O_3 in the sample = 0·696 ($V–v$).

B. Method for the Analysis of Lead Bisilicate Frits

PRINCIPLE OF THE METHOD

Determination of Loss on Ignition

As the material melts below 1000°C it is customary to carry out the determination of loss on ignition at a lower temperature, viz. 600°C.

Determination of Silica, Titania, Alumina and Lead Monoxide

The sample for the main part of the analysis is fused in sodium carbonate and boric acid, using a higher flux ratio than usual so that fusion may be achieved at a lower temperature and so minimising the risk of damage to the platinum dish by the lead in the sample. Nitric acid is used in preference to hydrochloric acid for dissolving the melt to avoid precipitation of lead chloride. After a few minutes evaporation to form a gel the silica is coagulated with polyethylene oxide, separated by filtration and ignited and weighed before and after treatment with hydrofluoric and sulphuric acids. The silica remaining in the filtrate is subsequently determined by a spectrophotometric method based on the formation of silicomolybdic acid, with or without reduction to molybdenum-blue.

The residue from the silica is fused in sodium carbonate and the solution of the melt added to the filtrate from the silica. After dilution to a standard volume titania is determined spectrophotometrically with hydrogen peroxide.

Alumina is determined volumetrically. After removal of lead by solvent extraction with sodium diethyldithiocarbamate (DDC)/chloroform followed by a cupferron/chloroform extraction of iron and titanium, excess EDTA is added, the alumina complexed by adjustment of the pH and, after boiling, the surplus EDTA is back-titrated with zinc solution using dithizone as indicator.

Lead monoxide is determined gravimetrically, after its removal from the DDC–chloroform phase with diluted nitric acid. The pH is adjusted and the lead precipitated as chromate in a buffered acetate solution.

Determination of Ferric Oxide, Lime, Magnesia and Alkalis

The sample is decomposed with hydrofluoric, nitric and sulphuric acids to remove the silica. A further evaporation with nitric and sulphuric acids removes traces of fluoride; the residue is dissolved in nitric acid and the solution diluted to a standard volume. Ferric oxide is determined spectro-photometrically with 1:10-phenanthroline hydrate. Lime and magnesia are determined volumetrically with EDTA, aliquots being withdrawn, dimerca-prol (BAL) added to complex the lead followed by triethanolamine to form complexes with the oxides of the ammonia group. The solutions are made strongly alkaline and then titrated for lime, and lime plus magnesia, screened Calcein and Solochrome Black 6B respectively being used as indicators. Alkalis are determined flame-photometrically.

REAGENTS

Additional Reagents

Ammonium acetate

B.A.L. (*2:3 Dimercapto-propan-1-ol*): Pass CO_2 into the bottle after use and stopper tightly. Store in a dark bottle.

Prepared Reagents

Ammonium acetate (*approx. 10%*): Dilute 140 ml of acetic acid (glacial) to

2000 ml with water and add carefully 140 ml of ammonia solution ($d = 0.88$). Mix and adjust to about pH 6.

Ammonium acetate buffer: Add 120 ml of acetic acid (glacial) to 500 ml of water followed by 74 ml of ammonia solution ($d = 0.88$). Dilute to 1 litre.

Ammonium molybdate (80 g/litre): Dissolve 80 g of ammonium molybdate in 1 litre of water and filter if necessary. Store in a polythene bottle and discard after 4 weeks or earlier if any appreciable deposit is observed.

BAL (7 + 93): Dilute 7 ml of BAL to 100 ml with ethanol (95%).

Cupferron (60 g/litre): Dissolve 3 g of cupferron in 50 ml of cold water; filter if necessary. This solution must be freshly prepared. If the reagent is discoloured or gives a strongly coloured solution a new stock should be obtained. The solid reagent should be stored in a tightly stoppered bottle in the presence of a piece of ammonium carbonate to prevent decomposition.

1:10-Phenanthroline hydrate (10 g/litre): Prepare enough solution for immediate use at a concentration of 0.1 g of 1:10-phenanthroline hydrate in 10 ml of acetic acid (1 + 1).

Polyethylene oxide (2.5 g/litre): Add 0.5 g of polyethylene oxide* to 200 ml of water slowly with stirring, preferably on a mechanical stirrer, until dissolved. Discard after 2 weeks.

Sodium diethyldithiocarbamate (DDC) (100 g/litre): Dissolve 5 g of sodium diethyldithiocarbamate in 50 ml of water and filter. This solution must be freshly prepared.

Sulphuric–nitric acid mixture: To 650 ml of water add 100 ml of diluted sulphuric acid (1 + 1) and 250 ml of nitric acid ($d = 1.42$).

Indicators

Bromophenol Blue
Calcein (screened)
Dithizone
2: 4-Dinitrophenol
Solochrome Black 6B.

Standard Solutions

EDTA (0.5M)
EDTA (5 g/litre)
Zinc (0.05M)

BLANK DETERMINATIONS

A blank determination should be carried out on all reagents in accordance with the general scheme of analysis. The fluxes should be weighed and transferred to the platinum dish and then dissolved by the addition of nitric acid,

---

*Union Carbide Polyox resins WSR 35, WSR N-80, WSR 205, WSR N-750 or WSR N-3000 are suitable as sources of polyethylene oxide.

the fusion being omitted to avoid undue attack on the platinum dish, which occurs in the absence of the sample.

The sample prepared for analysis should be ground to pass completely through a 120-mesh B.S. test sieve. A non-metallic (e.g. nylon bolting-cloth) sieve is preferable.

Determination of Loss on Ignition

Weigh 1·000 g of the prepared, dried (110°C) sample into a porcelain crucible. Raise the temperature slowly and finally ignite at 600°C for 30 min with free access to air to maintain oxidizing conditions.

Determination of Silica, Titania, Alumina and Lead Monoxide

Decomposition of the Sample

As lead is present in the sample extreme care must be taken during the fusion to guard against any reduction to metal that will then alloy with the platinum and ruin the dish. As an additional precaution the fusion should be carried out at as low a temperature as possible.

Weigh 1·000 g of the prepared, dried (110°C) sample into a platinum dish; a convenient size is 70-mm dia., 40-mm deep, of effective capacity 75 ml. Add 5 g of anhydrous sodium carbonate and 0·6 g of boric acid and mix thoroughly. Cover the dish with a lid.

Heat over a gas burner, cautiously at first until frothing ceases, then complete the fusion at as low a temperature and in as short a time as possible to ensure a clear melt, with occasional swirling of the dish and its contents to promote thorough mixing. Allow the dish and contents to cool.

Add 30 ml of diluted nitric acid (1 + 1) and a few drops of hydrogen peroxide (6%), to decompose any lead peroxide formed, and transfer the dish to a steam bath to facilitate dissolution of the melt. If the solution gels before dissolution is complete it may be necessary to stir the gel gently with a glass rod. When dissolution is complete, remove the lid and wash any spray adhering to it into the dish with the minimum amount of water.

Determination of the Main Silica

Allow the dish to remain on the steam bath until a stiff gel is formed, then thoroughly mix a Whatman accelerator tablet into the gel until the mixture is homogeneous, ensuring that the gel adhering to the sides of the dish is also broken up.

Add slowly, with stirring, 5 ml of polyethylene oxide solution (2·5 g/litre) and mix well, ensuring that any gel adhering to the sides of the dish is brought into contact with the coagulant. Add 10 ml of water, mix and allow to stand for 5 min.

Filter through a 125-mm No. 42 Whatman paper, transferring the silica to the filter with hot water, scrubbing the dish with a rubber tipped glass rod.

Wash the precipitate thoroughly with hot water. The silica precipitate obtained by this method is more voluminous than that obtained by dehydration and so care must be taken to ensure that the mass of precipitate is broken up thoroughly during the washing; this is particularly important here where it is imperative that no lead shall be left in the silica.

Transfer the paper and precipitate to an ignited and weighed platinum crucible. Ignite at a low temperature until the precipitate is free from carbonaceous matter and then heat in a furnace at 1200°C to constant weight, 30 min, usually being sufficient. (The ignition should be started at a lower temperature than normal or a noticeable amount of carbon may remain after ignition. In many cases a trace of carbon may remain but its weight appears to be negligible and does not affect the results.)

Moisten the contents of the cold crucible with water, add 5 drops of sulphuric acid (1 + 1) and 10 ml of hydrofluoric acid (40% w/w). Evaporate to dryness on a sand bath in a fume cupboard. For the evaporation, the crucible and contents should be heated from below. The use of top heating alone, as with a radiant heater, can result in incomplete elimination of silica by the hydrofluoric acid.

Heat the crucible and residue over a gas flame and finally ignite for 5 min at 1000°C, cool and weigh. If the residue weighs more than 5 mg, repeat the treatment with sulphuric and hydrofluoric acids to ensure that all the silica is removed. The difference between the two weights represents the "gravimetric" silica.

Preparation of the Solution for the Determination of Residual Silica, Titania, Alumina and Lead Monoxide

Fuse the residue from the hydrofluoric acid treatment of the "gravimetric" silica with 1 g of anhydrous sodium carbonate, cool and transfer the crucible and contents, with the lid, into the beaker containing the main filtrate. When the melt has dissolved, remove the crucible and lid and wash them thoroughly with hot water, scrubbing them with a bobby. Cool and dilute the solution to 500 ml in a volumetric flask. This solution is referred to as the "stock" solution.

Determination of Residual Silica

Silicomolybdate yellow method

Transfer a 20-ml aliquot of the stock solution to each of two 50-ml volumetric flasks, A and B, and add a few drops of 2: 4-dinitrophenol indicator. To flask B add ammonia solution ($d = 0.88$) dropwise until the indicator turns yellow (note the number of drops used) then add 5 ml of hydrochloric acid (1 + 4). To flask A add 5 ml of hydrochloric acid (1 + 4) and the same number of drops of ammonia as needed to neutralize the solution in flask B, then add 5 ml of ammonium molybdate solution (80 g/litre).

Dilute the solution in each flask to 50 ml and shake well. Measure the silicomolybdate colour in flask A against the solution in flask B in 40-mm cells at 440 nm or by using a colour filter (Ilford 601) in a suitable instrument.

Measure the optical density not sooner than 5 min and not later than 15 min after the addition of the ammonium molybdate solution.

Determine the silica content of the solution by reference to a calibration graph.

If a cloudy solution develops in carrying out the above method, the molybdenum-blue procedure should be used.

Molybdenum-blue method

Transfer 5 ml of the stock solution to a 100-ml volumetric flask A, add 15 ml of water and 2 drops of 2:4-dinitrophenol indicator solution. Add diluted ammonia solution (1 + 1) dropwise until the indicator turns yellow; noting the amount of ammonia used, then add 5 ml of diluted hydrochloric acid (1 + 4).

To another 100-ml volumetric flask B, add 20 ml of water and the same amount of diluted ammonia solution (1 + 1) as used to neutralize the aliquot in flask A. Add 2 drops of 2:4-dinitrophenol indicator followed by diluted hydrochloric acid (1 + 4) until the solution is neutral and then 5 ml in excess.

To both flasks add 6 ml of ammonium molybdate (80 g/litre) and stand for 5–10 min at a temperature not less than 20°C and not greater than 30°C. Then add, with swirling, 45 ml of diluted hydrochloric acid (1 + 1) and stand for 10 min.

Add 10 ml of stannous chloride (10 g/litre), dilute to 100 ml and mix. (The appearance of a brown colour in the solution is normal and does not interfere with the determination). Measure the optical density of the solution in flask A against the solution in flask B in 10-mm cells at 800 nm or by using a colour filter (Ilford 609) in a suitable instrument. The colour is stable between 5 and 30 min after the addition of the stannous chloride solution.

Calculation of the Total Silica Content

Add the residual silica content to the figure obtained for the "gravimetric" silica to obtain the total silica content.

Determination of Titania

Transfer 20 ml aliquots of the stock solution to each of two 50-ml volumetric flasks, A and B. To flask A only add 10 ml of hydrogen peroxide solution (6%).

Dilute the solution in each flask to 50 ml and shake well. Measure the optical density of the pertitanic acid solution in flask A against the solution in flask B in 40-mm cells at 398 nm, or by using a colour filter (Ilford 601) in a suitable instrument. The colour is stable between 5 min and 24 h after the addition of the hydrogen peroxide solution. Determine the titania content of the solution by reference to a calibration graph.

Determination of Alumina and Lead Monoxide

Separation of lead monoxide

Transfer a 200 ml aliquot of the stock solution to a 500-ml separating funnel (Squibb's type) A. Add 25 ml of chloroform and 20 ml of sodium diethyldithiocarbamate solution (100 g/litre). Stopper the funnel and shake vigorously for 1 min. Release the pressure in the funnel by carefully removing the stopper and rinse the stopper and neck of the funnel with water. Allow the layers to separate, then run the chloroform layer into another separating funnel B. Using a polythene wash bottle, rinse the stem of funnel A inside and out, with chloroform and allow the rinsings to run into funnel B.

To the solution in funnel A, add 10 ml of chloroform and 5 ml of sodium diethyldithiocarbamate solution (100 g/litre). Shake for 1 min, run the chloroform layer into funnel B and wash the funnel stem as before. Wash the aqueous solution in funnel A twice with 10-ml portions of chloroform and transfer the chloroform layers to funnel B. Reserve this chloroform solution for the determination of lead monoxide, and the aqueous solution in funnel A for the determination of alumina.

Determination of alumina

Separation of ferric oxide and titania. To the aqueous solution in funnel A, add 40 ml of hydrochloric acid ($d = 1\cdot18$), 20 ml of chloroform and 5 ml of cupferron solution (60 g/litre). Stopper the funnel and shake vigorously. Release the pressure in the funnel by carefully removing the stopper and rinse the stopper and neck of the funnel with water. Allow the layers to separate and withdraw the chloroform layer. Wash the aqueous solution twice with 10-ml portions of chloroform. Discard the chloroform extracts and wash the stem of the separating funnel, inside and out, with chloroform.

Volumetric determination of alumina. Transfer the aqueous solution to a 500-ml conical flask. Add a few drops of bromophenol blue indicator and then add ammonia solution ($d = 0\cdot88$), drop by drop, until the solution is just alkaline. Re-acidify quickly with hydrochloric acid ($d = 1\cdot18$) and add 5-6 drops in excess. Add 10 ml of standard EDTA solution ($0\cdot05\text{M}$) and then add ammonium acetate buffer solution until the indicator turns blue followed by 10 ml in excess. Boil the solution down to 150-200 ml and cool to room temperature.

Add a volume of ethanol (95%) equal to the volume of the solution followed by 1-2 ml of dithizone indicator and titrate with standard zinc solution ($0\cdot05\text{M}$) from blue-green to the first appearance of a permanent pink colour.

Calculation—If the EDTA solution is not exactly $0\cdot05\text{M}$, calculate the equivalent volume of exactly $0\cdot05\text{M}$ EDTA.

If V ml is the volume of EDTA ($0\cdot05\text{M}$) and v ml is the volume of zinc solution ($0\cdot05\text{M}$) used in back-titration, then Al_2O_3 (%) $= 0\cdot6375\,(V-v)$.

Extraction of lead monoxide from the organic phase

To the chloroform solution in separating funnel B add 50 ml of nitric acid ($2 + 3$). Stopper the funnel and shake vigorously for 30 sec. Release the pressure in the funnel by carefully removing the stopper and rinse the stopper and neck of the funnel with water. Allow the layers to separate and run the chloroform layer into separating funnel A. Retain the aqueous solution in funnel B.

To funnel A add 20 ml of diluted nitric acid (2 + 3) and shake vigorously. Discard the chloroform layer and transfer the aqueous solution to funnel B, rinsing funnel A with water and adding the rinsings to funnel B. Wash this solution three times with 10-ml portions of chloroform. Discard the chloroform extracts and wash the stem of the funnel inside and out, with chloroform.

Gravimetric determination of lead monoxide

Transfer the aqueous solution to a 400-ml beaker. Add a few drops of bromophenol blue indicator followed by ammonia solution ($d = 0.88$) until the indicator just turns blue. Add 3 ml of nitric acid ($d = 1.42$). Boil off all traces of chloroform.

Add 10 ml of potassium dichromate solution (50 g/litre) and 20 ml of ammonium acetate solution (250 g/litre). Allow the solution to cool and stand overnight. Filter through a weighed, sintered-glass crucible of porosity grade 4, scrubbing out the beaker with a bobby. Wash the precipitate thoroughly with cold water. Dry the crucible and contents at 110°C for 2 h, cool and weigh.

Calculation—Weight of $PbCrO_4 \times 0.6906 =$ weight of PbO.

Determination of Ferric Oxide, Lime, Magnesia and Alkalis

Decomposition of the Sample

Weigh 0.250 g of the prepared, dried (110°C) sample into a small platinum basin. Add 10 ml of sulphuric–nitric acid mixture and about 10 ml of hydrofluoric acid (40% w/w).

Cover the basin with a lid and digest the mixture in a fume cupboard on a hot sand bath to facilitate attack of the sample before evaporation; then evaporate to dryness, being careful to avoid spurting. Cool, add 10 ml of sulphuric–nitric acid mixture and rinse down the sides of the basin with water. Again evaporate carefully to dryness.

To the cool, dry residue add 20 ml of diluted nitric acid (1 + 19) and digest the basin and contents on a steam bath for 10 min. Cool, filter if necessary, and dilute the solution to 250 ml in a volumetric flask. This will be referred to as the "sample solution A".

NOTE: As sulphuric acid is used for the decomposition it may be preferable to determine lime and magnesia on the main stock solution.

Determination of Ferric Oxide

Transfer a 25-ml aliquot of the sample solution A to a 100-ml volumetric flask. Add 2 ml of hydroxyammonium chloride solution (100 g/litre), 5 ml of 1:10-phenanthroline solution (10 g/litre) and ammonium acetate solution (approx. 10%) until a pink colour appears in the solution, followed by 2 ml in excess.

Allow to stand for 15 min, dilute to 100 ml and shake well. Measure the optical density of the solution against water in 40-mm cells at 510 nm, or by using a colour filter (Ilford 603) in a suitable instrument. The colour is stable between 5 min and 75 min after the addition of the ammonium acetate solution. Determine the ferric oxide content of the solution by reference to a calibration graph.

Determination of Lime

Transfer an 80 ml aliquot of the sample solution A to a 500 ml conical flask. Add 2 ml of BAL solution (7 + 93). Mix together 10 ml of potassium hydroxide solution (250 g/litre) and 5 ml of triethanolamine (1 + 1) and add the mixture to the contents of the flask. Add about 0·015 g of screened Calcein indicator and titrate with standard EDTA solution (5 g/litre) from a 10 ml semi-micro burette, the colour change being from fluorescent green to pink.

Determination of the Sum of Lime and Magnesia

Transfer an 80 ml aliquot of the sample solution A to a 500 ml conical flask. Add 10 drops of hydrochloric acid ($d = 1·18$) and 2 ml of BAL solution (7 + 93). Mix together 25 ml of ammonia solution ($d = 0·88$) and 10 ml of triethanolamine (1 + 1) and add the mixture to the contents of the flask. Add about 0·07 g of Solochrome Black 6B indicator and titrate with standard EDTA (5 g/litre) from a 10-ml semi-micro burette, the colour change being from wine red to the last change to a clear blue.

Calculation of Magnesia

Subtract the volume of EDTA used for the titration of lime from the volume of EDTA used for the titration of the sum of lime and magnesia. The remainder represents the volume of EDTA required for the titration of magnesia.

Determination of Alkalis

Determine the alkali content on the remainder of the sample solution A used for the determination of ferric oxide, lime and magnesia. The procedure is described in Chapter 16 under "Lead Bisilicate Frit".

C. General Method for Colourless or White Opacified Glazes

GENERAL

The method described in this section is based fairly closely on Classical procedures, but includes some more modern techniques where these have been proved by experience. Section D outlines a number of possibilities for more rapid procedures, all of which have been used in the laboratories of the Association, but which have not been thoroughly evaluated.

Several variants of the method for the main analysis are described, the choice of procedure depending on the additional elements present. At first sight this may appear to make the description complicated, but if the simplest method for each set of circumstances is to be described, this is inevitable.

The following additional elements are dealt with in this chapter—lead, tin, barium, zirconium and zinc. No satisfactory wet chemical method has yet been found for the separation of barium, strontium and calcium when present together.

Materials which contain only the constituents normally found in aluminosilicates, with the addition of borate, may be analysed by the method described for borax frits (p. 156), and lead borosilicates may be analysed by the method for lead bisilicates (p. 165) but also determining borate as in the analysis of borax frits (p. 165).

REAGENTS

Additional Reagents:

Ammonium acetate
Ammonium sulphate
BAL (2: 3-dimercapto-propan-1-ol): Pass CO_2 into the bottle after use and stopper tightly. Store in a dark bottle.
Citric acid
Formic acid: 98–100% w/w
Mannitol
Sodium sulphide.

Prepared Reagents

Ammonium acetate (approx. 10%): Dilute 140 ml of acetic acid (glacial) to 2000 ml with water and add carefully 140 ml of ammonia solution ($d = 0.88$). Mix, cool and adjust to approximately pH 6, either with acetic acid or ammonia solution. Alternatively, this solution may be made by diluting ammonia acetate (approx. 40% — pH 6·0) with three times its volume of water.

Ammonium acetate buffer: Add 120 ml of acetic acid (glacial) to 500 ml of water followed by 74 ml of ammonia solution ($d = 0.88$). Mix, cool and dilute to 1 litre.

Ammonium nitrate (10 g/litre): Dilute 10 ml of nitric acid ($d = 1.42$) to about 200 ml. Add diluted ammonia solution (1 + 1) until the solution is faintly alkaline to methyl red. Dilute to 1 litre.

Ammonium phosphate (100 g/litre): Dissolve 2 g of di-ammonium hydrogen orthophosphate in 20 ml of water. This solution should be freshly prepared.

BAL (2: 3-dimercapto-propan-1-ol) (7 + 93): Add 7 ml of BAL to 93 ml of ethanol (95%) and mix. Store in a dark bottle.

Bromine water (saturated): Shake 20 ml of bromine with 500 ml of water in a glass-stoppered bottle.

Cupferron (60 g/litre): Dissolve 3 g of cupferron in 50 ml of cold water and filter. This solution must be freshly prepared. If the reagent is discoloured or gives a strongly coloured solution a new stock should be obtained. The solid reagent should be stored in a tightly-stoppered bottle in the presence of a piece of ammonium carbonate to prevent decomposition.

Cupferron wash liquor: Add 100 ml of hydrochloric acid ($d = 1·18$) to 800 ml of water. Dissolve 1·5 g of cupferron in 100 ml of cold water and filter. Mix the two solutions. This wash liquor should be freshly prepared.

Formic acid mixture: Dissolve 250 g of ammonium sulphate in about 500 ml of water, add 200 ml of formic acid (98–100% w/w) and 30 ml of ammonia solution ($d = 0·88$). Cool and dilute to 1 litre.

Formic acid wash liquor: Add 4 ml of formic acid (98–100% w/w) to 996 ml of water and saturate with hydrogen sulphide.

Hydrogen sulphide wash solution: Add 20 ml of hydrochloric acid ($d = 1·18$) to 980 ml of water and saturate with hydrogen sulphide.

1:10-Phenanthroline hydrate (10 g/litre): Prepare enough solution for immediate use at a concentration of 0·1 g of 1:10-phenanthroline hydrate in 10 ml of diluted acetic acid (1 + 1).

Silver nitrate solution (1 g/litre)—used for testing for chlorides; Dissolve 0·1 g of silver nitrate in 100 ml of diluted nitric acid (1 + 9).

Sodium sulphide solution (10 g/litre): Dissolve 1 g of sodium sulphide in 100 ml of water. This solution must be freshly prepared.

Sulphuric–nitric acid mixture: To 650 ml of water, add 100 ml of diluted sulphuric acid (1 + 1) and 250 ml of nitric acid ($d = 1·42$).

Indicators

Bromophenol blue
Calcein (screened)
Dithizone
Methyl red
Phenolphthalein
Solochrome Black 6B.

Standard Solutions

EDTA (0·05M)
EDTA (5 g/litre)
Hydrochloric acid (approx. 0·1N)
Potassium permanganate (approx. 0·1N)
Sodium hydroxide (approx. 0·1N)
Zinc (0·05M)

PREPARATION OF THE SAMPLE

The sample prepared for analysis should be ground to pass completely through a 120-mesh B.S. test sieve. A non-metallic (e.g. nylon bolting-cloth) sieve is preferable.

Most frits and glazes are already available in a finely ground condition as a "slop". This should be dried on a steam bath and the solid rubbed down in a porcelain or agate mortar.

Where the frit is in lump form this may generally be gently tamped in a porcelain mortar and finally ground in an agate mortar.

PROCEDURE

Determination of Loss on Ignition

It is difficult to specify a temperature to which frits and glazes should be ignited, as they contain relatively volatile elements and have different softening temperatures. The choice of ignition temperature is often a compromise. Carbon dioxide and water must be completely removed and as some glazes may contain calcium carbonate this must be decomposed. At the same time it is important not to exceed the softening temperature of the glaze, otherwise it may melt and prevent all the gases being released. Finally, it is imperative not to drive off zinc, lead, boron or alkali metals by volatilization.

Frits such as lead bisilicate are best ignited at about 600°C, whereas glazes containing calcium carbonate require a temperature of about 900°C.

Weigh 1·000 g of the finely ground, dried sample into a porcelain crucible. Heat over a mushroom burner, cautiously at first, then slowly raise the temperature to 600–900°C, depending on the nature of the sample. Heat at the chosen temperature for 30 min. Cool and weigh.

Decomposition of the Sample

WHENEVER LEAD OR TIN IS PRESENT IN THE SAMPLE, EXTREME CARE MUST BE TAKEN DURING THE FUSION TO GUARD AGAINST ANY REDUCTION TO METAL WHICH MAY THEN ALLOY WITH THE PLATINUM AND RUIN THE CRUCIBLE. As an additional precaution the fusion should be carried out at as low a temperature and for as short a time as possible.

Weigh 1·000 g of the finely ground, dried (110°C) sample into a platinum crucible, add 7 g of anhydrous sodium carbonate and mix thoroughly.

Heat over a mushroom burner, raising the temperature slowly until frothing ceases, then complete the fusion at as low a temperature and for as short a time as possible, occasionally swirling the melt to ensure thorough mixing. Quench the melt by immersing the bottom portion of the crucible in cold water, then place the crucible and lid in about 100 ml of water in a 250 ml beaker. Cover the beaker with a clock glass and introduce 20 ml of hydrochloric acid ($d = 1·18$) through the lip of the beaker. Warm until the melt is completely disintegrated and remove the crucible and lid, washing them thoroughly and scrubbing them with a "bobby".

If the melt can be detached from the crucible after quenching, the main bulk may be placed directly into a porcelain basin and the remainder of the melt removed from the crucible with a few millilitres of diluted hydrochloric acid (1 + 1). The main bulk of the melt is then dissolved in diluted hydrochloric acid in situ, as described above, and the solutions combined in the basin.

If the dissolution has been carried out in a beaker, transfer the solution to a porcelain basin, washing the beaker thoroughly and scrubbing out with a "bobby".

Crush any lumps remaining in the solution. If borate is known or suspected to be present, cool the solution slightly and add 25 ml of methanol to ensure the elimination of the borate as methyl borate.

Determination of Silica

Carefully evaporate the solution until the bulk of the alcohol has been removed and then evaporate to dryness on a steam bath, breaking up from time to time the crust that forms and hinders evaporation.

When the residue is completely dry, cover the basin with a clock glass and drench the residue with 10 ml of hydrochloric acid ($d = 1 \cdot 18$). Allow to stand for a few minutes and then add about 75 ml of hot water and digest on a steam bath for 5 min to dissolve the salts.

Filter through a No. 40 Whatman paper, transferring the silica to the filter with a jet of hot, diluted hydrochloric acid (1 + 19) and scrubbing the basin with a "bobby". Wash the residue five times with hot diluted hydrochloric acid (1 + 19) and then with hot water until it is free from chlorides.

Reserve the paper and residue for the subsequent ignition and transfer the filtrate and washings back to the evaporating basin. Add a further 25 ml of methanol and again evaporate to dryness on a steam bath. Cover the basin with a clock glass and bake in an air oven at 110°C for 1 h.

Cool, drench the residue with 10 ml of hydrochloric acid ($d = 1 \cdot 18$) and allow to stand for a few minutes. Then add about 75 ml of hot water and digest on a steam bath for 5 min to dissolve the salts.

Filter through a No. 42 Whatman paper, transferring the residue to the filter with a jet of hot, diluted hydrochloric acid (1 + 19) and scrubbing the basin with a "bobby". Wash five times with hot, diluted hydrochloric acid (1 + 19) and then with hot water until the residue is free from chlorides.

Reserve the filtrate and washings for the determination of lead (or R_2O_3). Place the two papers and residues, without drying, in a weighed platinum crucible and heat gently to dry the residues and char the papers. Then burn off the carbon at a low temperature, otherwise it may be impossible to remove all of it, and finally ignite at 1200°C for 30 min and then to constant weight. Allow to cool and weigh to obtain the weight of the impure silica.

In a number of cases the solution from the silica residue must not be combined with that from the silica, until after the determination of lead; hence it is essential that the separation of silica from lead should be carefully carried out.

Moisten the weighed residue with 10 drops of diluted sulphuric acid (1 + 1) and add about 10 ml of hydrofluoric acid (40% w/w). Evaporate to dryness on a sand bath in a fume cupboard.

Ignite the dry residue at 1200°C for 5 min, allow the crucible to cool and weigh. Subtract the weight of this residue from the weight of the impure silica to obtain the weight of the silica in the sample taken.

If the residue weighs more than about 5 mg, repeat the treatment with sulphuric and hydrofluoric acids to ensure that all the silica has been removed. If the residue is still in excess of 5 mg, the presence of an appreciable amount of barium, titanium or zirconium should be suspected.

Fuse the residue with about 1 g of potassium pyrosulphate and dissolve the melt in a little water to which about 2 ml of hydrochloric acid ($d = 1 \cdot 18$) has been added. Any barium in the residue will be precipitated as sulphate at this stage.

H

Hence, if metals of Group II are present the above solution is added to the filtrate and washings after the hydrogen sulphide precipitation.

If metals of Group II are not present the solution is added to the filtrate and washings from the silica determination.

Precipitation of Group II Metals

If metals of Group II are not present, barium is next determined (see p. 179).

To the filtrate and washings from the silica determination, add diluted ammonia solution (1 + 1), drop by drop, until the first appearance of a permanent precipitate. Just dissolve the precipitate by adding hydrochloric acid ($d = 1.18$) and then add 5 ml in excess.

Heat the solution to incipient boiling and then pass a slow stream of washed hydrogen sulphide through the solution until cold, and allow to stand overnight.

Filter the solution through a Buchner funnel with a sintered-glass mat (porosity 4) or a No. 42 Whatman paper (if tin is present a paper should be used), transferring the precipitate to the filter with a jet of cold hydrogen sulphide wash solution. Wash thoroughly with cold hydrogen sulphide wash solution. Reserve the filtrate and washings for the determination of barium (if present) and the ammonia group oxides.

Separation of Lead and Tin Sulphides

If tin is not present proceed to the determination of lead oxide (see below).

Transfer the paper and precipitate back to the precipitation beaker; dissolve about 1 g of sodium sulphide in about 50 ml of water, warm to 70°C, and pour into the beaker. Immediately cover the beaker with a clock glass and swirl vigorously for about 5 min. Macerate the filter paper with a glass rod to ensure that the tin sulphide is dissolved.

Filter through a No. 542 Whatman paper, transferring the pulp and residue to the filter with a jet of sodium sulphide solution (10 g/litre). Wash thoroughly with sodium sulphide solution and then three times with hydrogen sulphide wash solution.

Reserve the filtrate and washings for the determination of tin oxide. The lead sulphide remains on the filter and is determined as described below.

Determination of Lead Oxide

Dissolve the precipitate through the filter with 40 ml of hot, diluted hydrochloric acid (1 + 1) and wash thoroughly with hot water. If the separation from tin has been carried out, great care must be taken to see that all the lead is extracted from the paper pulp. Any platinum that may have dissolved as a result of attack on the crucible during the fusion will be precipitated with the sulphides and will remain on the filter.

Transfer the solution and washings to a 400-ml beaker (or back to the precipitation beaker if tin has not been separated). Evaporate carefully to dryness; on a hot plate until the volume is reduced to about 20 ml and then on a steam bath to avoid spurting.

Add 3 ml of hydrochloric acid ($d = 1·18$) and 100 ml of water. Heat to boiling and boil until all the lead chloride has dissolved. Add 5 ml (10 ml if the amount of lead is expected to exceed 30% PbO) of potassium dichromate solution (50 g/litre) and 20 ml of ammonium acetate solution (250 g/litre).

Allow the solution to cool and stand overnight. Filter through a weighed crucible with a sintered-glass mat (porosity 4) and wash eight times with cold water. Dry for 2 h at 110°C, cool and weigh.

Calculation—Weight of $PbCrO_4 \times 0·6906$ = weight of PbO.

Determination of Tin Oxide

Cover the beaker containing the solution reserved for the determination of tin oxide, with a clock glass and carefully introduce 20 ml of nitric acid ($d = 1·42$) through the lip of the beaker. When the vigorous reaction has ceased, wash and remove the clock glass and evaporate the solution on a steam bath to a volume of about 5 ml, but **not to dryness.**

Dilute to about 50 ml with hot water and add 15 ml of nitric acid ($d = 1·42$). Evaporate to about 5 ml as before. This digestion ensures the quantitative precipitation of tin as metastannic acid.

Dilute to about 75 ml with hot water, add a small amount of paper pulp and stand on a steam bath for 30 min, with occasional swirling.

Filter through a No. 42 Whatman paper; if the filtrate is not clear at first, pass it again through the filter and wash thoroughly with hot, diluted nitric acid ($1 + 99$).

Place the paper and precipitate in a weighed porcelain crucible and heat gently to dry the residue and char the paper. Burn off the carbon at as low a temperature as possible to avoid reduction of stannic oxide. Cool, moisten with nitric acid ($d = 1·42$), carefully drive off the excess acid and finally ignite at 1000°C for a few minutes. Cool and weigh as SnO_2.

Determination of Barium Oxide

If barium oxide is not present proceed to the determination of the ammonia group oxides (see p. 180).

Boil the solution reserved for the determination of barium oxide until all the hydrogen sulphide has been expelled. To the boiling solution add, drop by drop from a pipette, 10 ml of diluted sulphuric acid ($1 + 9$). Boil for a further few minutes and digest on a steam bath for 3 h.

Allow to cool and stand for at least 4 h, or preferably overnight.

Filter through a No. 42 Whatman paper, transferring the precipitate to the filter with a jet of hot water and scrubbing the beaker with a "bobby". Wash free from chlorides with hot water. Reserve the filtrate and washings for the determination of the ammonia group oxides.

Transfer the paper and precipitate to a weighed platinum crucible and heat gently to dry the residue and char the paper. Burn off the carbon at as low a temperature as possible and finally ignite at 1000°C for 30 min. Cool and weigh as barium sulphate.

Calculation—Weight of $BaSO_4 \times 0·6570$ = weight of BaO.

Determination of the Ammonia Group Oxides

If zinc is present or the Classical procedure is to be used for the determination of lime and magnesia, it is necessary to separate the ammonia group oxides by means of an ammonia precipitation.

If zinc is not present and the lime and magnesia are to be determined by the EDTA method then the ammonia group oxides can be determined directly on the solution reserved from the determination of barium.

Precipitation of the Ammonia Group Oxides

If zinc is present proceed as described below.

The method of precipitation of the ammonia group oxides depends on the amount present and, therefore, this section has been divided into parts (a) and (b). Part (a) deals with glazes and borax frits, which contain an appreciable amount of R_2O_3 and need three precipitations; part (b) deals with lead bisilicate frits, which contain less than 5% R_2O_3 and require only two precipitations.

Add 5 ml of bromine water to the filtrate and washings from the barium determination and boil for a few minutes to ensure that the iron is present in the ferric state. If no barium precipitation has been carried out, it will be necessary to boil off the excess of hydrogen sulphide, if Group II metals were present.

(a) More than about 5% R_2O_3.

To the solution (approx. 300 ml) add 2–3 g of ammonium chloride, warm the solution to about 80°C and add diluted ammonia solution (1 + 1), with stirring, until the solution is just alkaline to bromophenol blue. Boil the alkaline solution for 2 min, allow to stand for 5 min for the precipitate to settle, and filter through a No. 41 Whatman paper.

Transfer the precipitate to the filter with a jet of hot, faintly ammoniacal ammonium nitrate solution (10 g/litre) and wash five times with hot, faintly ammoniacal ammonium nitrate solution (10 g/litre).

Just acidify the filtrate and washings with hydrochloric acid ($d = 1·18$) and reserve; transfer the precipitate and paper back to the precipitation beaker. Dissolve the precipitate in a slight excess of hydrochloric acid ($d = 1·18$) and macerate the filter paper. Dilute to about 250 ml and repeat the precipitation as before.

Filter through a No. 541 Whatman paper, transferring the precipitate to the filter with a jet of hot, faintly ammoniacal ammonium nitrate solution (10 g/litre) and scrubbing the beaker with a "bobby".

Wash the precipitate free from chlorides with hot faintly ammoniacal ammonium nitrate solution (10 g/litre). Just acidify the filtrate and washings with hydrochloric acid ($d = 1·18$) and add to the reserved solution. Retain the precipitate and paper for the subsequent ignition.

Evaporate the combined filtrates from the two precipitations to about 150 ml and add diluted ammonia solution (1 + 1), with stirring, until the solution is just alkaline to bromophenol blue.

Boil off the excess ammonia, allow to stand for 5 min and filter through a

No. 40 Whatman paper. Transfer the precipitate to the filter with a jet of hot faintly ammoniacal ammonium nitrate (10 g/litre) scrubbing the beaker with a "bobby", and wash free from chlorides with hot, faintly ammoniacal ammonium nitrate solution (10 g/litre).

Just acidify the filtrate and washings with hydrochloric acid ($d = 1 \cdot 18$) and reserve for the determination of zinc. Reserve the paper and precipitate for subsequent ignition.

(b) *Less than about 5% R_2O_3.*

To the solution (approx. 200 ml) add 2–3 g of ammonium chloride, warm the solution to about 80°C and add diluted ammonia solution (1 + 1), with stirring, until the solution is just alkaline to bromophenol blue. Boil off the excess ammonia, allow to stand for 5 min for the precipitate to settle, and filter through a No. 40 Whatman paper. Transfer the precipitate to the filter with a jet of hot, faintly ammoniacal ammonium nitrate solution (10 g/litre) and wash five times with hot, faintly ammoniacal ammonium nitrate solution (10 g/litre).

Just acidify the filtrate and washings with hydrochloric acid ($d = 1 \cdot 18$) and reserve for the determination of zinc.

Transfer the precipitate and paper back to the precipitation beaker and dissolve the precipitate in a slight excess of hydrochloric acid ($d = 1 \cdot 18$). Macerate the filter paper.

Dilute to about 150 ml and repeat the precipitation as before. Filter through a No. 540 Whatman paper, transferring the precipitate to the filter with a jet of hot, faintly ammoniacal ammonium nitrate solution (10 g/litre) and scrubbing the beaker with a "bobby".

Wash free from chlorides with hot, faintly ammoniacal ammonium nitrate solution (10 g/litre). Just acidify the filtrate and washings with hydrochloric acid ($d = 1 \cdot 18$) and add to the solution reserved for the determination of zinc. Reserve the paper and precipitate for the subsequent ignition.

Place the paper and precipitate reserved for the ignition in a weighed platinum crucible and heat gently to dry the precipitate and char the paper. Burn off the carbon at a low temperature and finally ignite at 1200°C to constant weight; 30 min is usually sufficient. The weight of the mixed ferric, titanium and aluminium (also zirconium if present) oxides is thus obtained.

Carefully fuse the ignited, weighed oxides in the same platinum crucible with 10 g of potassium pyrosulphate. Cool the crucible and extract the melt in a 250-ml beaker containing about 150 ml of water to which 10 ml of sulphuric acid ($d = 1 \cdot 84$) had been cautiously added.

Cool the solution, transfer to a 500-ml volumetric flask, dilute to 500 ml and mix.

Separation of the Ammonia Group Oxides

The solution for the separation of the ammonia group oxides may be the solution from the potassium pyrosulphate fusion, the filtrate from the silica determination, the filtrate from the separation of the Group II metals or the filtrate from the determination of barium, depending on the constituents found in the material.

If one of the above filtrates is to be used, cool and transfer the filtrate and washings to a 500-ml volumetric flask, dilute to 500 ml and mix. Any of these solutions, including the solution from the potassium pyrosulphate fusion, will be referred to as the "stock test solution".

The solution may contain only ferric oxide, titania and alumina or it may contain zirconia in addition. The presence of zirconia will necessitate slight modifications to the methods for the determination of titania and alumina, and the zirconia will itself need to be determined. If a considerable amount of zirconia is present it may prevent the detection of end-points in the EDTA titrations of lime and magnesia, in which case these elements will need to be determined in the solution which has been reserved for the determination of zinc.

Determination of Ferric Oxide

In some cases the iron content of frits and glazes may be critical and at the same time may be present in very small quantities. In these cases, e.g. lead bisilicate frits, it is advisable to take the aliquot for the determination of iron from the solution prepared for the determination of alkalis, thus avoiding large blanks such as may result from the addition of the large number of reagents necessary to reach this point in the analysis.

Transfer a 5-ml aliquot of the stock test solution or the solution prepared by the hydrofluoric acid decomposition to a 100-ml volumetric flask.

Add 2 ml of hydroxyammonium chloride solution (100 g/litre), 5 ml of 1:10-phenanthroline solution (10 g/litre) and 2 ml of ammonium acetate solution (approx. 10 %).

Allow to stand for 15 min, dilute to 100 ml and mix.

Measure the optical density of the solution against water in 10-mm cells at 510 nm or by using a colour filter (Ilford 603) in a suitable instrument. The colour is stable between 15 min and 75 min after the addition of the ammonium acetate solution.

Determine the ferric oxide content of the solution by reference to a calibration graph.

Determination of Titania

Transfer a 20-ml aliquot of the stock test solution to each of two 50-ml volumetric flasks, A and B. To each flask add 10 ml of diluted phosphoric acid (2 + 3) (If zirconia is known to be present the phosphoric acid is omitted.) and, to flask A only, 10 ml of hydrogen peroxide solution (6 %). Dilute the solution in each flask to 50 ml and shake well.

Measure A against B in 40-mm cells at 398 nm or by using a colour filter (Ilford 601) in a suitable instrument. The colour is stable between 5 min and 24 h after the addition of the hydrogen peroxide solution.

Determine the titania content of the solution by reference to a calibration graph.

Determination of Zirconia

Transfer a 100-ml aliquot of the stock test solution to a 400-ml beaker and add 20 ml of hydrochloric acid ($d = 1.18$). Cool the solution by immersing the bottom half of the beaker in running water.

Add 20 ml of cupferron solution (60 g/litre), with stirring, and a small amount of filter-paper pulp (ashless). Stand the beaker and contents in the cold for 5 min and add a few more drops of cupferron solution (60 g/litre) to check that no further precipitation takes place.

Filter the precipitate through a No. 42 Whatman paper, transferring the precipitate to the filter with a jet of cold cupferron wash liquor and scrubbing the beaker with a "bobby". Wash thoroughly with cold cupferron wash liquor.

It is essential to keep the solution cold during filtration and washing, otherwise the precipitate tends to become tarry and cannot be effectively washed.

Transfer the precipitate and paper to a weighed platinum crucible and discard the filtrate and washings. Heat gently to dry the precipitate and char the paper; great care must be taken at this stage as the precipitate itself tends to melt and some may be easily lost by spurting.

Burn off the carbon at a low temperature and finally ignite at 1200°C to constant weight; 30 min is usually sufficient. Cool and weigh. The weight is that of zirconia, ferric oxide and titania in the aliquot taken.

Calculation—Calculate the weight of the mixed oxides in the 500 ml and deduct the weight of ferric oxide and titania from the total weight to obtain the weight of zirconia in the sample taken.

Determination of Alumina

Separation of ferric oxide, zirconia (if present) and titania

Transfer a 100-ml aliquot of the stock test solution to a 500-ml separating funnel. If a separation of ammonia group oxides has been carried out add 20 ml of hydrochloric acid ($d = 1.18$); otherwise, add 10 ml of hydrochloric acid ($d = 1.18$) and 25 ml of diluted sulphuric acid $(1 + 1)$.

Add 20 ml of chloroform and an appropriate amount of cupferron solution (60 g/litre). For glazes which do not contain zirconia or added titania 5 ml will be sufficient, but if zirconia is present or if titania is present in large amounts, it will be necessary to add cupferron solution until the addition of a few more drops does not produce a permanent precipitate.

Stopper the funnel and shake vigorously. Release the pressure in the funnel by carefully removing the stopper and rinse the stopper and neck of the funnel with water.

Allow the layers to separate and withdraw the chloroform layer. Confirm that extraction is complete by checking that the addition of a few drops of cupferron solution (60 g/litre) to the aqueous solution does not produce a permanent precipitate.

Add further 10-ml portions of chloroform and repeat the extraction until the chloroform layer is colourless. Discard the chloroform extracts.

Determination of alumina

Run the aqueous solution from the cupferron-chloroform extraction into a

500-ml conical flask. Add a few drops of bromophenol blue indicator; then add ammonia solution ($d = 0 \cdot 88$) until the solution is just alkaline.

Re-acidify quickly with hydrochloric acid ($d = 1 \cdot 18$) and add 5–6 drops in excess.

Add sufficient standard EDTA solution (0·05M) to provide an excess of a few millilitres over the expected amount. Then add ammonium acetate buffer solution until the indicator turns blue, followed by 10 ml in excess. Boil the solution for 10 min and cool rapidly.

Add an equal volume of ethanol (95%) and 1–2 ml of dithizone indicator, and titrate with standard zinc solution (0·05M) from blue-green to the first appearance of a permanent pink colour.

Calculation—If the EDTA solution is not exactly 0·05M calculate the equivalent volume of 0·05M EDTA.

If V ml is the volume of EDTA (0·05M) and

v ml is the volume of zinc solution (0·05M) used in the back-titration, then the percentage of alumina, x, is given by:

$$x = 1 \cdot 275\,(V - v)$$

Determination of Zinc Oxide

Adjust the volume of the filtrate and washings from the ammonia group precipitation to 250–300 ml and transfer to a 500-ml conical flask.

Add ammonia solution ($d = 0 \cdot 88$) to the change point of bromophenol blue and then add 25 ml of citric acid solution (200 g/litre) and again adjust the pH to the change point of bromophenol blue with ammonia solution ($d = 0 \cdot 88$).

Add 25 ml of formic acid mixture, heat to about 90°C and insert a two-hole stopper carrying an outlet tube flush with the bottom of the stopper and an inlet tube extending nearly to the liquid. Pass a stream of washed hydrogen sulphide and displace the air.

Close the outlet tube and allow the solution to cool while it is saturated with hydrogen sulphide under increasing pressure. Shake occasionally and allow to stand for about 1 h.

Filter through a Buchner funnel with a sintered-glass mat (porosity 4) or a No. 42 Whatman paper, transferring the precipitate to the filter with a jet of cold formic acid wash liquor. Reserve the filtrate and washings for the determination of lime.

Dissolve the precipitate through the filter with 20 ml of diluted hydrochloric acid (1 + 1) and wash thoroughly with hot water. Return the solution and washings to the precipitation flask and boil to expel hydrogen sulphide.

Cool, neutralize the solution with ammonia solution ($d = 0 \cdot 88$) to the change point of bromophenol blue, re-acidify quickly with hydrochloric acid ($d = 1 \cdot 18$) and add 5–6 drops in excess.

Add sufficient standard EDTA solution (0·05M) to provide an excess of a few millilitres over the expected amount. Then add ammonium acetate buffer solution until the indicator turns blue and then add 10 ml in excess.

Add an equal volume of ethanol (95%) and 1–2 ml of dithizone indicator, and titrate with standard zinc solution (0·05M) from blue-green to the first appearance of a permanent pink colour.

Calculation—If the EDTA solution is not exactly 0·05M calculate the equivalent amount of 0·05M EDTA.

If V ml is the volume of EDTA (0·05M) and
  v ml is the volume of zinc solution (0·05M) used in the back-titration,
  then the percentage of ZnO, x, is given by:

$$x = 0·4069 \ (V - v)$$

Determination of Lime, Magnesia and Alkalis

Decomposition of the Sample

Weigh 0·250 g of the finely ground, dried (110°C) sample into a small platinum basin.

Add 10 ml of sulphuric–nitric acid mixture and about 10 ml of hydrofluoric acid (40% w/w).

Transfer the vessel to a sand bath, allow to react thoroughly with the lid on for about 15 min, then evaporate to dryness in a fume cupboard, being careful to avoid spurting.

Cool, add 10 ml of sulphuric–nitric acid mixture and rinse down the sides of the basin with water. Evaporate carefully to dryness.

To the cool, dry residue add 20 ml of diluted nitric acid (1 + 19) and digest on a steam bath for 10 min.

Cool, filter if necessary, dilute the solution to 250 ml in a volumetric flask and mix to form the sample solution A.

Determination of Lime and Magnesia: EDTA Method

Unless the sample contains a large amount of zirconia it is possible to carry out the determinations volumetrically by EDTA on aliquots of the solution prepared for the determination of alkalis. As this method is much simpler than the Classical, it is always advisable to make the attempt (see p. 53) before proceeding with the classical determinations.

Determination of lime

If lead, tin and zinc are known to be absent, the BAL which is used to complex these elements is unnecessary and the method may be used exactly as described in the Coagulation method for aluminosilicates.

Transfer a 80-ml aliquot of the sample solution A prepared by the attack of hydrofluoric acid to a 500-ml conical flask. Add 1 ml of BAL solution or, if the material is a lead bisilicate frit, add 2 ml. Mix together 10 ml of potassium hydroxide solution (250 g/litre) and 5 ml of diluted triethanolamine (1 + 1) and add the mixture to the contents of the 500-ml conical flask. Add about 0·015 g of screened Calcein indicator and titrate with standard EDTA solution (5 g/litre) from a semi-micro burette, the colour change being from fluorescent green to pink.

H*

Determination of the sum of lime and magnesia

Transfer a 80-ml aliquot of the sample solution A prepared by the attack of hydrofluoric acid to a 500-ml conical flask. Add 20 drops of hydrochloric acid ($d = 1.18$). Add 1 ml of BAL solution or, if the material is a lead bisilicate frit, add 2 ml. Mix together 25 ml of ammonia solution ($d = 0.88$) and 10 ml of diluted triethanolamine (1 + 1), and add the mixture to the contents of the flask. Add about 0.07 g of Solochrome Black 6B indicator and titrate with standard EDTA solution (5 g/litre) from a semi-micro burette, the colour change being from wine red to clear blue.

Calculation of magnesia

The volume of EDTA used for the titration of the lime is subtracted from the volume of EDTA used for the titration of the sum of lime and magnesia.

The remainder represents the volume required for the titration of magnesia.

Determination of Alkalis

Determine the alkali content on an aliquot of the sample solution A by the procedure described in Chapter 16 under "Glazes".

Determination of Lime and Magnesia: Classical Method

If the end-points for the titrations of lime and magnesia with EDTA are obscured by the presence of zirconium, which generally precipitates, it is necessary to carry out the determinations of lime and magnesia by classical methods.

Boil down the filtrate and washings from the determination of zinc to 200–250 ml, expelling hydrogen sulphide.

Cool, add 20 ml of diluted sulphuric acid (1 + 1) and 100 ml of nitric acid ($d = 1.42$). Cover the beaker with a clock glass and evaporate carefully to fumes. If the solution shows a tendency to become syrupy and frothy, cool and add an additional volume of nitric acid. Repeat the evaporation, to remove the citric acid etc., until the solution is free of organic matter while strong fumes of sulphur trioxide are evolved.

Determination of lime

Cool, carefully add 20 ml of water and 3 g of oxalic acid. Boil the solution, then add with stirring, diluted ammonia solution (1 + 1) until the solution is alkaline to bromophenol blue, and then add an excess of 10 ml of diluted ammonia solution (1 + 1). Cover the beaker with a clock glass and digest on a steam bath for 2 h.

Allow to cool and stand for 1 h. Filter through a Buchner funnel with a sintered-glass mat (porosity 4) or a No. 42 Whatman paper, transferring the precipitate to the filter with a jet of cold ammonium oxalate solution (10 g/litre). Wash the precipitate four times with cold ammonium oxalate solution (10 g/litre).

Reserve the filtrate and washings for the determination of magnesia. Dissolve the precipitate through the filter with 20 ml of hot, diluted nitric acid (1 + 1) and wash the filter thoroughly with hot water.

Transfer the solution and washings back to the precipitation beaker, add 1 g of oxalic acid, boil the solution and precipitate the calcium as before.

Allow to cool and stand for at least 1 h. Filter through a Buchner funnel with a sintered-glass mat (porosity 4) or a No. 42 Whatman paper, transferring the precipitate to the filter with a jet of cold water. Wash the precipitate thoroughly with cold water. Add the filtrate and washings to the solution reserved for the determination of magnesia and acidify with hydrochloric acid ($d = 1 \cdot 18$).

Dissolve the precipitate through the filter with 50 ml of hot, diluted sulphuric acid (1 + 9) and wash the filter thoroughly with hot water. Transfer the solution back to the precipitation beaker, heat to boiling and titrate with standard potassium permanganate solution (approx. $0 \cdot 1 \text{N}$) until the first appearance of a permanent pink colour.

Calculation—1 ml $0 \cdot 1 \text{N}$ $KMnO_4$ ≡ $0 \cdot 002804$ g CaO.

Determination of magnesia

Evaporate the acidified solution reserved for the determination of magnesia to about 300 ml. Add 10 ml of ammonium phosphate solution (100 g/litre) and heat to boiling. Make just alkaline to bromophenol blue with diluted ammonia solution (1 + 1) and cool to room temperature. Add 20 ml of ammonia solution ($d = 0 \cdot 88$), stir vigorously to start the precipitation and allow to stand overnight at a low temperature.

Filter through a No. 42 Whatman paper, transferring the precipitate to the filter with a jet of cold, diluted ammonia solution (1 + 39), and wash four times with cold, diluted ammonia solution (1 + 39). Discard the filtrate and washings.

Dissolve the precipitate through the filter with hot, diluted hydrochloric acid (1 + 1) and wash thoroughly with hot water. Transfer the solution and washings back to the precipitation beaker. Add 1 ml of ammonium phosphate solution (100 g/litre), heat the solution to boiling and, while stirring, make just alkaline to bromophenol blue with diluted ammonia solution (1 + 1).

Cool, add 10 ml of ammonia solution ($d = 0 \cdot 88$), and stir vigorously to start the precipitation. Allow to stand overnight at a low temperature.

Filter through a No. 42 Whatman paper, transferring the precipitate to the paper with a jet of cold, diluted ammonia solution (1 + 39) and scrubbing the beaker with a "bobby". Wash thoroughly with cold, diluted ammonia solution (1 + 39), discarding the filtrate and washings.

Transfer the precipitate and paper to a weighed platinum crucible and heat gently to dry the precipitate and char the paper. Burn off the carbon at a low temperature. (Platinum crucibles may be seriously attacked during the ignition if the precipitate is not thoroughly washed with diluted ammonia solution, or if the precipitate is ignited too strongly before all the carbon is oxidized.)

Finally ignite at $1150°C$ for 15 min, cool and weigh the residue as magnesium pyrophosphate.

Calculation—Weight of $Mg_2P_2O_7 \times 0\cdot3623$ = weight of MgO.

Determination of Boric Oxide

Boric oxide is determined by the method described under 'A' Analysis of Borax Frits, p. 165, **bearing in mind that whenever lead or tin is present in the material great care must be exercised during the fusion to prevent any reduction to metal which may then alloy with the platinum and ruin the crucible.**

D. Possible More Rapid Procedures

GENERAL

Many of the procedures described in C are time-consuming and offer difficulties in manipulation. Some of the elements which may be present can cause considerable difficulties in some mixtures. Zinc, for example, is very prone to precipitation in the ammonia group and can be extremely difficult to separate. On the other hand, the formic acid precipitation technique is equally difficult to handle, great care, and in some cases, a little luck being required to ensure a pure precipitate even after a reprecipitation.

It is therefore possible, on a number of occasions to use less well-tried procedures with an equal or better chance of success. However, many of these procedures have merely been used on one or two occasions by the Association's laboratory, with no reason to doubt the accuracy of the final result. Because of this it seems sufficient to describe these merely as possible methods in outline, rather than to attempt to give full working detail.

Most of the additions of elements such as tin, barium, zirconium and zinc, are of the order of 5% content, as is strontium on the rare occasion when this is used. Many glazes have lead contents in the range $7-10\%$, with similar levels of lime and alumina, magnesia is in the $0\cdot01-0\cdot1\%$ range. The possibilities of atomic absorption spectrophotometry and to a lesser extent flame photometry should be borne in mind as the former technique would appear eminently suitable for most of these elements.

Silica

On a number of occasions it has been found possible to dispense with double dehydration, substituting a coagulation. This, of course, leaves about 1% of silica in solution, but this is not always a problem. One way in which this has been overcome is by fusing 2 g of sample in a platinum dish; as there is no need for a dehydration stage when polyethylene oxide coagulant is used the amount of salts introduced is no handicap. The main silica is then separated by filtration and after adding the solution of the silica residue to the silica filtrate, the combined solution is made up to volume. Aliquots can be taken from this stock solution for various determinations, the actual elements depending on the mixture to be analysed. Residual silica in solution, iron and titanium are almost certainly worth doing on aliquots of this solution by the usual colorimetric methods.

It may even pay to use the coagulation in preference to double dehydration even when it may involve confirming possible silica contamination of, say, a zirconia precipitate obtained with cupferron; the balance of advantage in terms of time needs to be considered but more especially the reliability of the final result.

Tin

On occasion this has been removed from aliquots with cupferron together with iron and titania, but most frequently lead is also present which makes cupferron separation difficult, the use of sulphuric acid to increase the acidity being inadmissible, and even hydrochloric acid gives rise to difficulty in the cold conditions used due to the relative insolubility of lead chloride.

Fortunately the use of tin in glazes is diminishing, the combination of zinc and zirconia being preferred.

Lead

In addition to the gravimetric method described in Section B, there is a volumetric variant which may be of value in a number of instances. The chromate gravimetric method in Section B yields slightly discoloured pre-cipitates and the results appear to be slightly high ($\approx 0 \cdot 1 \%$), but the volu-metric method requires more manipulation as the organic solvent is lighter than water. However, the use of D.D.C. and its retention in the recovered lead aqueous phase prevent the use of a convenient volumetric finish. The method utilised a perchloric acid dissolution of the melt in place of nitric acid and a $0 \cdot 4$-g aliquot to which 8 ml of hydrochloric acid ($d = 1 \cdot 18$) had been added in a total volume of 200 ml. 20 ml of potassium iodide solution (250 g/litre) is added and a 50 ml of a $(1 + 4)$ isobutyl methyl ketone solution of Amberlite LA-1 liquid ion-exchange resin used for a solvent extraction. A further 25 ml of $(1 + 4)$ resin and washing with ketone bring the lead into the organic phase.

The organic phase is treated with 100 ml of NaOH (50 g/litre) and 50 ml of ethanol to minimise emulsification followed by a further 50 ml of NaOH and 50 ml of ethanol, thus returning the lead to the aqueous phase. After boiling to expel the ketone $2 \cdot 5$ g of sodium tartate is added and the pH adjusted with HCl to the change point of 2: 4-dinitrophenol. 30 ml of ammonia/ ammonium chloride buffer (35 g NH_4Cl + 212 ml ammonia [$d = 0 \cdot 88$] to 500 ml) is added and the lead titrated with EDTA using thymolphthalein complexone as indicator. It is better, when possible, to add the bulk of the EDTA before the neutralization.

It is possible to carry out the determination of alumina on the aqueous phase from the resin extraction after removal of iron and titanium with cup-ferron.

When zinc is present in addition to lead the D.D.C. extraction technique as described is not applicable as the two elements are not separated. According to the literature it should be possible either to separate the elements by D.D.C. extraction at different acidities or to re-extract them into the aqueous phase at different acid strengths.

In the presence of barium, it is a moot point whether there is a greater danger of losing barium in precipitating lead as sulphide or separating by solvent extraction with D.D.C. The other elements present would normally decide this issue, the presence of zinc necessitating sulphide precipitation and zirconium, with the consequent danger of hydrolysis, practically precluding this course.

Barium

Barium must be determined in the main analysis by precipitation as the sulphate, but the presence of considerable amounts of calcium in most glazes leads to the danger of co-precipitation of calcium sulphate. The use of sulphamic acid as precipitant is claimed to reduce this danger.

Where barium has to be removed in order to carry out other determinations the gravimetric determination offers the most economical approach. However, it is possible to determine barium by the difference in volumes of EDTA used in separate aliquots, one of which has been treated with sulphuric acid and the precipitate filtered off. The barium titrates with the calcium when Calcein is used as indicator, as also does strontium. The method offers little advantage in practice, since the conversion factor for barium is very unfavourable; the main advantage is that if both methods of carrying out the determination are used it is usually possible to detect the deliberate addition of strontium owing to a discrepancy in the barium results.

Barium and strontium, however, may well best be determined by flame spectrophotometry or atomic absorption spectrophotometry.

Zirconium

Zirconium is normally determined by cupferron, but clearly mandelic acid or its derivatives have some advantages in specificity. However, it has been shown feasible (Chapter 13) to remove all the zirconia in a 1-g sample by a cupferron precipitation, destroy the organic colouring matter and then use the solution for the determination of the remaining constituents. It has not proved possible to do this with any of the mandelic acids. The use of a 1-g aliquot of sample permits an accurate determination of the zirconia content.

Zinc

There are several possible methods whereby the zinc content may be arrived at including two approaches using an EDTA finish. The first is carried out on the main analysis solution and is best performed after fusing a two gram sample as suggested under "silica". The method consists basically in carrying out two determinations of alumina, the first normally with a cupferron/chloroform separation and the second with a D.D.C. separation and the acidity then increased for the cupferron/chloroform separation. In the first portion the iron and titanium are removed and the second zinc, iron and titanium, and thus zinc and aluminium are titrated in the first and aluminium only in the second, allowing zinc to be determined by difference.

A similar technique involves taking a 0·5-g sample for the determination of alkalis, lime and magnesia, carrying out the normal calcium, and calcium + magnesium determinations using BAL to complex the zinc; then if zinc is the only heavy metal present (i.e. no Pb or Sn), a similar titration for calcium + magnesium but omitting the BAL will also titrate the zinc thus enabling its content to be found.

Comparative results have been made on a number of samples by both these techniques with very good agreement.

Advantage has also been taken of the selective non-precipitation of aluminium by 8-hydroxyquinaldine to separate zinc and aluminium. The precipitating conditions for aluminium oxinate appear to be satisfactory for the precipitation of zinc with 8-hydroxyquinaldine so that using acetic acid/ammonium acetate buffer the zinc is precipitated with 8-hydroxyquinaldine, filtered off, dried and weighed and the aluminium then precipitated by the addition of 8-hydroxyquinoline, filtered off, dried and weighed.

CHAPTER 16

DETERMINATION OF ALKALIS

GENERAL

The most commonly used procedure for the determination of alkalis relies on flame photometry and, in the UK, the most commonly used instrument is the Evans Electroselenium Model 100. This equipment has the advantage of simplicity, being designed for a coal-gas flame and using filters for selecting wavelength. However, its supremacy in the field is now being challenged by other, more expensive, instruments. Evans Electroselenium Ltd. have recently introduced two new instruments, the Model 170 and the Model 227, both using the same atomiser—or, as it is now termed, nebulizer—and filter system. Both instruments have internal standardization by lithium, the 170 has a digital readout and the 227 integrates the signal over a 15-second period and the total is then read off on meters whose needles are stationary.

In addition to these, competition is growing from emission modifications to atomic absorption spectrophotometers. Most of these instruments either have or can be modified to have, flame emission capability and as they can burn propane and have monochromators they may well have advantages. Errors resulting, for example, from leakage of calcium light through a sodium filter are reduced and the increase in sophistication may yield steadier readings.

The method for the determination of alkalis described in the earlier editions of this book depended for its success on the ability to compare simple standards of the alkali sulphates dissolved in water with the alkalis derived from a prepared sample solution. The validity of this comparison was established almost 20 years ago in the laboratories of the B.Ceram.R.A. but the work was only published to member-firms and not released to any of the analytical journals. Since then a number of minor changes have been made in the instrument, principally in the nebulizer system which is now more coarse than in early models and, probably more significantly, coal gas has become a thing of the past. Town gas has for some years been variable and will shortly become almost exclusively natural gas, which has a very high percentage of methane. An extensive investigation carried out revealed that the original premise was no longer valid and a number of interferences were encountered.

In view of the variability of town gas supplies and as the amount of interference varied with gas composition it was decided to recommend the use of a bottle gas. The choice of propane for this purpose was determined by the fact that the composition of bottled propane is reputed to be more consistent than other hydrocarbons and the interference levels were found in general, to be lower than with town gas or methane. A new burner was designed to burn propane satisfactorily (Fig. 21) and one or two minor modifications made to the instrument to increase the ease of use. These included replacing the present potentiometer with a multi-turn helical potentiometer (20 K ohm) and connecting a capacitor (250 μF) across the terminals of the photocell.

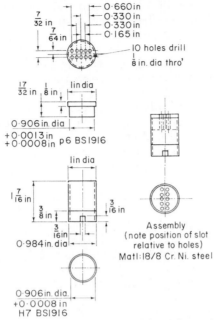

Fig. 21. Propane burner for EEL Model 100 flame photometer.

With this arrangement and burning propane it was found that both chloride and perchlorate depressed the readings whereas nitrate and sulphate did not. The chloride concentration can be controlled by the amount of hydrochloric acid added and its effect balanced by adding the same amount to the standards but the perchlorate is the amount left behind after evaporation to dryness. This will depend on the temperature of the sand bath, ensuring complete elimination of free perchloric acid (which can be controlled) and the amount retained as the perchlorates of the cations such as the alkalis and aluminium (which cannot be controlled). These facts demonstrated the need for a modified decomposition procedure using nitric and sulphuric acids.

Two forms of spectral interference were also noted, both of which required remedial action. Calcium light was again found to leak through the sodium filter and, to a lesser extent, the lithium filter. The addition of aluminium sulphate was found to overcome this, but a greater amount than previously recommended was found necessary. Finally it was found that the presence of potassium enhanced the sodium emission. This was accounted for on the basis of the relative ease of ionization. The flame causes dissociation of the salts into atoms and establishes equilibria as follows:

$$\text{Na (atom)} \rightleftharpoons \text{Na}^+\text{(ion)} + e$$
$$\text{K (atom)} \rightleftharpoons \text{K}^+\text{(ion)} + e$$

As the ease of ionization increases in the order Li, Na, K, Rb, Cs, it follows that potassium will produce a relatively greater proportion of ions and therefore electrons. These electrons will force the equilibrium in the first equation to the left and thus increase the concentration of sodium atoms in the flame. As the sodium atom is responsible for the production of light, this situation will increase the sodium light output.

This difficulty was overcome by the addition of the even more readily ionizable caesium into the solution, thus holding the sodium atom population to the maximum by introducing an overwhelmingly large supply of electrons into the flame.

Thus the method most recently recommended involves hydrofluoric, nitric and sulphuric acid decomposition, removal of fluorides with nitric and sulphuric acids and dissolution of the residue in nitric acid. After diluting to 250 ml, an aliquot of 25 ml is taken and caesium and aluminium sulphates are added so that after dilution to 50 ml the solution is 30 ppm with respect to caesium and 200 ppm with respect to the added Al_2O_3. This solution is compared to full-scale and zero solutions containing the same caesium and aluminium concentrations. In order to avoid further dilutions in most cases the top standard now contains 20 ppm K_2O, and 5 ppm Na_2O (as sulphates) giving equivalences of 4 % K_2O and 1 % Na_2O in the sample.

Dilutions may be made either by carrying out a preliminary dilution of the sample solution or by diluting an aliquot of the solution to which caesium and aluminium have been added with the zero standard solution thus maintaining the concentration of additives.

The Lawrence Smith method for the determination of alkalis has been omitted from the present volume since it now appears to be obsolete. Gravimetric determinations are used in some cases, usually for high contents of alkalis, for example, in the glass industry. The method for the determination of sodium is described under the method for the analysis of borax frit (p. 164), although in the light of the latest investigations it is doubtful whether equally good results could not be obtained using flame photometry. There is considerable disagreement concerning the correct drying of the sodium zinc uranyl acetate precipitate. Most procedures follow the practice of Barber and Kolthoff (*J. Am. chem. Soc.* **50**, 1625, 1928), of drying at room temperature. Vogel ("Quantitative Inorganic Analysis" (1966), Longmans) proposes drying at 55–60°C with the alternative of room temperature. Erdey ("Gravimetric Analysis", Part II (1965), Pergamon) on the other hand shows thermogravimetrically that a drying temperature of 105°C is to be preferred giving a precipitate of the composition $Na.Zn (UO_2)_3 (C_2H_3O_2)_9$ 6·5 H_2O. He also quotes results by Rády showing that a positive error of about 1 % occurs after drying by air suction compared with only about 0·2–0·3 % if the precipitate is dried at 105°C.

Potassium may be determined gravimetrically or volumetrically after precipitating with tetraphenylboron. For details of a suitable procedure reference may be made to that published by the Chemical Analysis Committee of the Society of Glass Technology (*Glass Tech.* **10**, 6, (1969)).

REAGENTS

Additional Reagents

Caesium sulphate (Johnson–Matthey "Specpure")

Prepared Reagents

Aluminium sulphate (10 mg Al_2O_3/ml approx.): Clean pure aluminium metal with hydrochloric acid, ethanol and ether. Weigh 10·58 g of the clean, dry metal, add 40 ml of sulphuric acid ($d = 1·84$), 120 ml of nitric acid ($d = 1·42$) and approximately 50 ml of water. Allow to react in the cold, raise the temperature gradually and heat until all the metal has dissolved and nitric oxides are expelled and the sulphuric acid fumes strongly. Cool and dissolve the crystalline melt in distilled water and dilute to 2 litres.

NOTE: The alkali content of Analar aluminium sulphate is too high to allow this to be used for the preparation of the above solution.

Aluminium sulphate (2000 ppm Al_2O_3 approx.): Dilute 400 ml of aluminium sulphate (10 mg Al_2O_3/ml approx.) to 2 litres.

Caesium sulphate (300 ppm Cs approx.): Dissolve 0·82 g caesium sulphate (Johnson Matthey "Specpure") in water and dilute to 2 litres.

Sulphuric–nitric acid mixture: To 650 ml of water, add 100 ml diluted sulphuric acid (1 + 1) and 250 ml of nitric acid ($d = 1·42$).

Standard Solutions

Lithium concentrated standard solution A ($Li_2O = 400$ ppm)
Lithium standard solution B ($Li_2O = 40$ ppm)
Potassium/sodium concentrated standard solution A ($K_2O = 400$ ppm, $Na_2O = 100$ ppm)
Potassium/sodium standard solution B ($K_2O = 40$ ppm, $Na_2O = 10$ ppm)

METHODS

The materials can be divided into two distinct types. The first type is that for which the sample decomposition has already been described in the appropriate chapter, since the solutions are also used for the determination of calcium and magnesium (i.e. aluminosilicates, aluminous and high-silica materials, frits, and glazes). The second type is that for which the decomposition and preparation of the sample solution have not been given (i.e. magnesites, dolomites, zircon-bearing materials, chrome-bearing materials and bone-ashes).

The determination of the alkalis in "soluble salts" is given separately on p. 201.

A. Procedure for the Determination of Alkalis in High-silica, Aluminosilicate, and Aluminous Materials, Borax Frits, Lead Bisilicate Frits, and Glazes (and Extension to Include High-lithium Materials)

INTRODUCTION

The sample decomposition for all the above materials is identical and is given in the appropriate Chapters. The resulting "sample solution A" is used for the determination of alkalis, lime and magnesia (and sometimes ferric oxide). As sulphuric acid is used in the decomposition it may be advisable to determine lime and magnesia on the main stock solution (prepared from the silica filtrate and silica residue) if this has also been prepared.

The method given below is essentially for the modified EEL Model 100 flame photometer, burning propane, where interference effects were extensively re-evaluated in 1970.

Sodium and potassium are determined on dilution of "sample solution A" after addition of caesium and aluminium sulphates. Lithium is determined directly on "sample solution A".

Samples should contain less than 10% CaO, i.e. less than 50 ppm CaO in the solution presented to the instrument, because the amount of aluminium sulphate added to overcome the spectral interference from lime on the sodium determination is insufficient for higher percentages.

Calibration has been adjusted to give full scale readings for 4% K_2O and 1% Na_2O in order to avoid dilutions for most materials other than fluxes, high alkali glazes, etc.

CALIBRATION

Additional Solutions

Lithium top standard (Li_2O = 20 ppm): Dilute 25·0 ml of the lithium concentrated standard solution A and 40 ml diluted nitric acid (1 + 19) to 500 ml in a volumetric flask and mix.

Potassium/sodium top standard (K_2O = 20 ppm, Na_2O = 5 ppm): To a 1-litre volumetric flask add 40 ml of diluted nitric acid (1 + 19), 100 ml of caesium sulphate solution (300 ppm Cs), 30 ml of aluminium sulphate solution (10 mg Al_2O_3/ml approx.) and 50·0 ml of the potassium/sodium concentrated standard solution A, dilute to 1 litre and mix.

Potassium/sodium zero standard (and diluent for alkalis in excess of 4% K_2O, 1% Na_2O): To a 1-litre volumetric flask, add 40 ml diluted nitric acid (1 + 19), 100 ml of caesium sulphate solution (300 ppm Cs) and 30 ml of aluminium sulphate solution (10 mg Al_2O_3/ml approx.), dilute to 1 litre and mix. Check the blanks for Na_2O and K_2O by spraying with the zero set on distilled water and the full scale set on the top standard. Typical values are deflections of 2 for Na_2O and 7 for K_2O. Readings greatly in excess of this would invalidate the use of the solution and the origin of the blank must be investigated and rectified.

Intermediate Calibration Standards

Lithium: To four 50-ml volumetric flasks, add 4 ml of diluted nitric acid (1 + 19) and 10, 20, 30 and 40 ml of the lithium top standard, dilute to 50 ml with water and mix. These concentrations are equivalent to 4, 8, 12 and 16 ppm Li_2O or 0·4, 0·8, 1·2 and 1·6% Li_2O.

Potassium/sodium: To seven 200-ml volumetric flasks add 8 ml of diluted nitric acid (1 + 19), 20 ml of the caesium sulphate solution (300 ppm Cs), 6 ml of aluminium sulphate solution (10 mg Al_2O_3/ml approx.) and 5, 10, 20, 40, 50, 60 and 80 ml respectively of the potassium/sodium standard solution B to give, 1, 2, 4, 8, 10, 12 and 16 ppm K_2O and 0·25, 0·5, 1, 2, 2·5, 3 and 4 ppm Na_2O respectively, equivalent at the dilution used to 0·2, 0·4, 0·8, 1·6, 2·0, 2·4 and 3·2% K_2O and 0·05, 0·1, 0·2, 0·4, 0·5, 0·6 and 0·8% Na_2O respectively.

Setting up the Instrument

About 15 min before the instrument is used the gas should be lit (by a taper as ordinary gas lighters will not ignite propane) and the air turned on; and adjusted to 10 lb/in². Spray water and adjust the gas flow to give well-defined blue cones; the gas mixture should not be set so lean that the flame "lifts-off" when the water is withdrawn. When the instrument has warmed up check the gas and air settings.

Calibration

Insert the appropriate filter.

For lithium set full scale on the 20 ppm Li_2O top standard and zero on distilled water. Spray the intermediate calibration standards and note the readings, checking full scale and zero frequently.

From the readings prepare a calibration graph.

For potassium and sodium set full scale on the potassium/sodium top standard and zero on the potassium/sodium standard. Spray the intermediate calibration zero standards and note the readings, checking full scale and zero frequently.

From the readings prepare a calibration graph.

Lithium is subject to positive errors caused by potassium light passing through the filter. The magnitude of the error must be ascertained by setting up as for the lithium calibration and spraying solutions of 20, 50 and 100 ppm K_2O equivalent to sample contents of 2, 5 and 10% K_2O and a correction graph prepared.

PROCEDURE

NOTE: The "sample solution A" is that prepared by hydrofluoric, nitric and sulphuric acid attack of the sample and is described in the appropriate Chapter.

Determination of Alkalis

Potassium and Sodium

Transfer 25·0 ml of the "sample solution A" to a 50-ml volumetric flask, containing 5 ml each of caesium sulphate solution (300 ppm Cs) and aluminium sulphate solution (2000 ppm Al_2O_3)*. Dilute with water to 50 ml to give "sample solution B".

*Attempts to simplify the procedure by making a joint solution of caesium and aluminium sulphates is not advised, as the slow precipitation of sparingly soluble caesium alum will cause serious loss of caesium.

Set up the instrument (as described previously), insert the appropriate filter and spray the "sample solution B", setting full scale on the potassium/sodium top standard and the zero with the potassium/sodium zero standard. If the alkali content is in excess of the normal range, readings can be brought on to scale, either by dilution of "sample solution B" with zero standard or by appropriate dilution of an aliquot of "sample solution A" before addition of caesium and aluminium sulphates. It is essential to maintain in the solutions presented to the instrument concentrations equivalent to additions of 5 ml of caesium sulphate solution (300 ppm Cs) and 5 ml of aluminium sulphate solution (2000 ppm Al_2O_3) in 50 ml of solution.]

Lithium

Insert the lithium filter and spray the "sample solution A", setting full scale on 20 ppm Li_2O and zero on distilled water. Note the reading and apply the appropriate correction for the K_2O content determined previously, by reference to the correction graph.

EXTENSION OF THE METHOD TO HIGH-LITHIUM MATERIALS

Introduction

Certain ceramic materials are very similar in composition to aluminosilicates, except that they contain larger quantities of lithium, e.g. petalite, lepidolite, amblygonite, etc. Glazes may also contain 2–3 % Li_2O.

The procedure adopted for the determination of alkalis in aluminosilicates, etc, previously described, can be applied directly to these high-lithium materials with the exception of the determination of lithium. This modification is described below.

Calibration for Lithium

Additional Solutions

High-lithium top standard (Li_2O = 20 ppm): To a 1-litre volumetric flask, add 40 ml of dilute nitric acid (1 + 19), 100 ml of caesium sulphate solution (300 ppm Cs), 30 ml of aluminium sulphate solution (10 mg Al_2O_3/ml approx.) and 50·0 ml of the lithium concentrated standard solution A, dilute to 1 litre and mix.

High-lithium zero standard: use the potassium/sodium zero standard as prepared on p. 196.

Intermediate calibration standards

To seven 200-ml volumetric flasks add 8 ml of dilute nitric acid (1 + 19), 20 ml of the caesium sulphate solution (300 ppm Cs), 6 ml of the aluminium sulphate solution (10 mg Al_2O_3/ml approx.) and 5, 10, 20, 40, 50, 60 and 80 ml of the lithium standard solution B to give 1, 2, 4, 8, 10, 12 and 16 ppm Li_2O respectively, equivalent at the dilution used to 0·2, 0·4, 0·8, 1·6, 2·0, 2·4 and 3·2 % Li_2O respectively.

Calibration

Insert the lithium filter and set up the instrument as previously described on p. 197.

Set full scale on the 20 ppm high-lithium top standard and zero on the high-lithium zero standard. Spray the intermediate calibration standards and note the readings, checking full scale and zero frequently. From the readings prepare a calibration graph.

Procedure

The aliquot of the "sample solution A" prepared for the determination of potassium and sodium by the addition of caesium and aluminium sulphates described on p. 197 is also used for the determination of lithium.

Insert the lithium filter in the instrument and spray the diluted aliquot setting full scale on the high-lithium top standard and zero on the high-lithium zero standard. Interpolate the % Li_2O from the calibration graph, prepared from the intermediate calibration standards. Correct the % Li_2O for the interference from potassium by reference to the correction graph (p. 197).

B. Procedure for the Determination of Alkalis in Magnesites, Dolomites, Zircon-bearing Materials, Chrome-bearing Materials and Bone-ashes

INTRODUCTION

This method is designed for materials with a low alkali content and where no other determination is to be made on the prepared solution.

The sample decomposition for all the above materials is identical to that given for aluminosilicates etc. except that only 0·1 g sample is used. The final volume is made up to 100 ml after the addition of appropriate amounts of caesium and aluminium sulphates. For this class of material the resultant solution is not used for the determination of any other constituent and so this technique allows sodium, potassium and lithium to be determined on one solution with full scale readings equivalent to 0·5% Na_2O, 2·0% K_2O and 2·0% Li_2O. The amount of aluminium sulphate added in this procedure is increased so that the final concentration is 3000 ppm Al_2O_3 in the solution presented to the instrument.

CALIBRATION

Additional Solutions

Lithium/sodium/potassium zero standard: To a 1-litre volumetric flask, add 200 ml of diluted nitric acid (1 + 19), 100 ml of caesium sulphate solution (300 ppm Cs) and 300 ml of aluminium sulphate solution (10 mg Al_2O_3/ml approx), dilute to 1 litre and mix.

Sodium/potassium top standard (K_2O = 20 ppm, Na_2O = 5 ppm): To a

1-litre volumetric flask add 200 ml of diluted nitric acid (1 + 19), 100 ml of caesium sulphate solution (300 ppm Cs), 300 ml of aluminium sulphate solution (10 mg Al_2O_3/ml approx.) and 50·0 ml of the potassium/sodium concentrated standard solution A, dilute to 1 litre and mix.

Lithium top standard (Li_2O = 20 ppm): To a 1-litre volumetric flask, add 200 ml of diluted nitric acid (1 + 19), 100 ml of caesium sulphate solution (300 ppm Cs), 300 ml of aluminium sulphate solution (10 mg Al_2O_3/ml approx.) and 50 ml of the lithium concentrated standard solution A, dilute to 1 litre and mix.

Calibration

It is not necessary to prepare calibration graphs specially for this procedure. The graphs already prepared for the aluminosilicate procedure may be used (p. 197). It should be noted, however, that the results for Na_2O and K_2O should be divided by a factor of two due to the fact that the solution presented to the instrument is twice the strength as that in the aluminosilicate procedure, but Li_2O results may be read direct from the graph.

PROCEDURE

Decomposition of the Sample

Weigh 0·100 g of the finely ground, dried (110°C) sample into a small platinum basin and ignite to remove organic matter if necessary.

To the cool dish add 10 ml of sulphuric-nitric acid mixture and about 10 ml of hydrofluoric acid (40% w/w). Transfer the vessel to a sand bath in a fume cupboard, allow to react thoroughly with the lid on for about 15 min, then evaporate to dryness, being careful to avoid spurting.

Cool, add 10 ml of the sulphuric–nitric acid mixture, and rinse down the sides of the basin with water. Evaporate carefully to dryness.

To the cool, dry residue add 20 ml of diluted nitric acid (1 + 19) and warm to dissolve as much of the residue as possible.

Transfer the cool solution, filtering if necessary through a washed paper, into a 100-ml volumetric flask containing 30 ml of aluminium sulphate solution (10 mg Al_2O_3/ml approx.) and 10 ml of caesium sulphate solution (300 ppm Cs). Dilute to 100 ml and mix to form the sample solution.

Determination of Alkalis

Insert the appropriate filter and set up the instrument as previously described.

Set full scale on the appropriate top standard and zero on the zero standard. Spray the sample solution and note the readings for K_2O, Na_2O and Li_2O.

Read the results from the graphs prepared for the aluminosilicate procedure. Divide sodium and potassium results by a factor of two, and correct the lithium result for the potassium in the sample by reference to the correction graph previously prepared (p. 197).

C. Procedure for the Determination of "Soluble Salt" Alkalis

INTRODUCTION

The preparation of the sample solution involves the extraction of the finely ground sample with water and is described in Chapter 24. Acids are absent, so that no interference from chlorides or perchlorates should be encountered. However, lime may be present in large quantities and a procedure involving the addition of aluminium and caesium sulphates is therefore necessary, in order to eliminate enhancement and spectral interference effects.

CALIBRATION

No special calibration graphs are necessary, the results for potassium and sodium can be obtained using the calibration graphs prepared for the aluminosilicate procedure (p. 197), after the necessary allowance for differences in sample concentration.

PROCEDURE

Transfer 25·0 ml of the soluble salt extract to a 50-ml volumetric flask containing 5 ml of caesium sulphate solution (300 ppm Cs) and 5 ml of aluminium sulphate solution (2000 ppm Al_2O_3), dilute with water to 50 ml and mix, to form the sample solution.

Set up the instrument as described previously and insert the appropriate filter. Set full scale on the potassium/sodium top standard and zero on the potassium/sodium zero standard (see p. 196). Spray the sample solution and note the reading. If this is more than full scale the reading may be brought back on to the scale either by dilution with zero standard of the sample solution, to which caesium and aluminium have been added, or alternatively by appropriate dilution of an aliquot of the soluble salt extract before addition of caesium and aluminium. It is essential to maintain, in the solutions presented to the instrument, concentrations equivalent to additions of 5 ml of caesium sulphate (300 ppm Cs) and 5 ml of aluminium sulphate (2000 ppm Al_2O_3) in 50 ml of solution.

Interpolate the results from the calibration graphs prepared in the aluminosilicate procedure (p. 197) and make the necessary allowance for difference in sample concentration.

CHAPTER 17

DETERMINATION OF ALUMINA USING DCTA

GENERAL

Almost all the methods outlined in the previous chapters include a determination of alumina with EDTA (ethylenediamine tetra-acetic acid) after separation of interfering elements by means of solvent extraction. These methods are in many cases standard procedures and, as such, have been allowed to remain in the descriptions of complete analytical methods. However, the technique has recently been improved by the substitution of DCTA (1: 2-diaminocyclohexane tetra-acetic acid). This reagent has the advantage over EDTA that alumina is complexed in the cold whereas with EDTA it is necessary to boil the solution for at least 10 minutes to ensure that all the alumina has been complexed. In terms of time alone, therefore, the analysis can be shortened by the length of time that it takes to bring the solution to boiling, 10 min boiling period and finally at least 10-15 min to cool down the solution again. Thus, at least 35–40 min can be saved.

DCTA also has the advantage that its reaction with chromium in the cold is very slow so that aluminium can be complexed and the excess of reagent back-titrated with zinc without interference from chromium. This, of course, has particular value in the analysis of basic refractories, magnesites and chrome-bearing materials, where chromium might normally be expected to be present.

The only problem arising from the substitution of DCTA for EDTA was occasional destruction of the indicator. It was found that this could be overcome by the addition of a reducing agent, hydroxyammonium chloride, to the solution prior to the indicator. This oxidation of the dithizone has been reported as being due to the use of chloroform; the boiling associated with the use of EDTA removes these organics.

Detailed instructions for the method are given below, together with an outline for its inclusion in the analytical procedures for various materials. The procedure is generally regarded as the more useful method, the extra cost of the reagent being more than recovered by the saving in time.

Aluminosilicates, Aluminous and High-silica Materials

The procedure may be applied as written as a direct substitute for the EDTA method.

Frits, Glazes

DCTA may be substituted for EDTA in these procedures. Preliminary separation will remain the same and DCTA procedure started at the point of the addition of the measured volume of reagent.

Magnesites and Dolomites

Chromium interferes stoicheiometrically in the determination of alumina with EDTA. For this reason it is necessary to carry out a determination of chromium in the sample and to ensure that all the chromium is precipitated with the aluminium in the R_2O_3 separation procedure. The latter, in turn, means that sulphurous acid needs to be added to reduce any Cr^{VI} to Cr^{III} and the excess removed by boiling, and as this process results in the iron in the sample being reduced to Fe^{++}, this must be oxidized by boiling the solution with nitric acid. This whole process adds at least half an hour to the determination. As an alternative, the reduction stage could be omitted but this would mean that a larger aliquot would have to be taken and a portion of this, after treatment, would have to be taken for the determination of chromium remaining in solution.

When DCTA is used these difficulties do not arise. No attempt need be made to reduce all the chromium, in fact, it may be advantageous to allow the bulk to escape, since any possible interference will be minimized. Thus the aliquot used for the determination of alumina (p. 109) may be neutralized directly with ammonia solution, the precipitate filtered off and redissolved as described and the DCTA method applied as below. It is essential, however, to ensure that the solution, after neutralization prior to the addition of DCTA, should be cold otherwise chromium may be complexed.

Chrome-bearing Materials

The method described in Chapter 12 incorporates the use of DCTA for the determination of alumina. The liquid ion-exchange resin separation can leave up to 0.5% Cr_2O_3 in the stock solution so that the use of DCTA is again preferable to EDTA.

REAGENTS

Prepared Reagents

Ammonium acetate buffer: Add, with stirring, 120 ml of acetic acid (glacial) to 500 ml of water, followed by 74 ml of ammonia solution ($d = 0.88$). Cool, dilute to 1 litre and mix.

Cupferron (60 g/litre): Dissolve 1·5 g of cupferron in 25 ml of water and filter. This solution must be freshly prepared.

Indicators

Bromophenol blue
Dithizone

Standard Solutions

DCTA (0·05M)
Zinc (0·05M)

PROCEDURE

(High-silica, aluminosilicate and aluminous materials).

NOTE: The "stock" solution is that obtained by combination of the silica filtrate and the solution of the silica residue, after removal of silica by the single dehydration or coagulation methods.

Separation of Ferric Oxide and Titania

For high-silica materials, transfer a 200-ml aliquot of the stock solution to a 500-ml separating funnel and add 40 ml of hydrochloric acid ($d = 1.18$). If the material is either known or suspected to contain a lime content of equal to or more than the alumina content, add 30 ml of hydrochloric acid ($d = 1.18$) and 25 ml of diluted sulphuric acid ($1 + 1$).

For aluminosilicates and aluminous materials, transfer a 100-ml aliquot of the stock solution to a 500-ml separating funnel and add 20 ml of hydrochloric acid ($d = 1.18$). If the material is suspected to contain more than about 5% of lime, add 10 ml of hydrochloric acid ($d = 1.18$) and 25 ml of diluted sulphuric acid ($1 + 1$).

To the appropriate solution add 20 ml of chloroform and 10 ml of cupferron solution (60 g/litre). Stopper the funnel and shake vigorously. Release the pressure in the funnel by carefully removing the stopper and rinse the stopper and neck of the funnel with water.

Allow the layers to separate and withdraw the chloroform layer. Confirm that extraction is complete by checking that the addition of a few drops of cupferron solution (60 g/litre) to the aqueous solution does not produce a permanent coloured precipitate.

Add further 10-ml portions of chloroform and repeat the extraction until the chloroform layer is colourless. Discard the chloroform extracts.

Run the aqueous solution from the cupferron-chloroform separation into a 1000-ml conical flask. Add a few drops of bromophenol blue indicator and add ammonia solution ($d = 0.88$) until just alkaline.

Re-acidify quickly with hydrochloric acid ($d = 1.18$) and add 5–6 drops in excess.

Add sufficient standard DCTA solution (0.05M) to provide an excess of a few millilitres over the expected amount. Then add ammonium acetate buffer solution until the indicator turns blue followed by 10 ml in excess.

Add an equal volume of ethanol (95%), 20 ml of hydroxyammonium chloride (100 g/litre) and 1–2 ml of dithizone indicator, and titrate with standard zinc solution (0.05M) from green to the first appearance of a permanent pink colour. (The end point is often improved by the addition of a little naphthol green B solution (1 g/litre) so as to eliminate any pink colour which may be formed in the solution on the addition of the indicator.)

Calculation—If the DCTA solution is not exactly 0.05M calculate the equivalent volume of 0.05M DCTA.

If V ml is the volume of DCTA (0.05M) and v ml is the volume of standard zinc solution (0.05M) used in the back-titration, then

$$Al_2O_3 \ (\%) = 1.275 \ (V-v) \text{ for a 100-ml aliquot and}$$
$$Al_2O_3 \ (\%) = 0.6375 \ (V-v) \text{ for a 200-ml aliquot.}$$

CHAPTER 18

DETERMINATION OF SULPHATE IN SILICATE MATERIALS

GENERAL

Two methods of decomposition of the sample are described, one using a sodium hydroxide fusion and the other a sodium carbonate fusion. The former is faster and is applicable to routine work on normal high-silica materials and aluminosilicates which are attacked by the flux. It cannot be used for materials such as aluminous refractories, or those which have relatively high contents of barium sulphate. The method using sodium carbonate should be retained for accurate work and for materials to which the first method is not applicable.

PRINCIPLE OF THE METHOD

The sample is decomposed by means of an alkaline fusion using either sodium carbonate or sodium hydroxide. The melt is leached with water and the residue filtered off. To the solution containing sulphate and other soluble anions, together with some silica and alumina, is added bromine to oxidize any sulphur to the sulphate state. The solution is then acidified and boiled to expel carbon dioxide.

Barium chloride is added to the boiling solution and the whole allowed to stand overnight. The precipitated barium sulphate is filtered off, ignited and weighed. For accurate work or when there is any indication of precipitated silica, such as slow filtration, the residue is treated with hydrofluoric and sulphuric acids, evaporated, ignited and again weighed.

REAGENTS

Prepared Reagents:

Bromine water (saturated): Shake 20 ml of bromine with 500 ml of water in a glass-stoppered bottle.

Silver nitrate (1 g/litre): Dissolve 0·1 g of silver nitrate in 100 ml of diluted nitric acid (1 + 9).

PROCEDURE

Decomposition of the Sample

Sodium Carbonate Fusion

Weigh 1·000 g of the finely ground, dried (110°C) sample into a platinum crucible and mix with about 7 g of anhydrous sodium carbonate. Transfer the crucible to a mushroom burner and slowly raise the temperature until frothing ceases, then complete the fusion at 1000°C for 15 min.

Quench the melt by immersing the bottom half of the crucible in cold water, then place the crucible and lid in about 100 ml of water contained in a 250-ml beaker. Cover the beaker with a clock glass and stand on a steam bath until the melt has disintegrated.

Remove the crucible and lid, transferring the residue to the beaker with a jet of hot water. Scrub the crucible and lid with a "bobby".

Filter through a No. 40 Whatman paper, transferring the precipitate to the filter with a jet of hot water and wash thoroughly with hot water.

Sodium Hydroxide Fusion

Carefully fuse 7 g of sodium hydroxide pellets in a nickel crucible (40 mm dia. and 40 mm deep) until any water is driven off and a clear melt is obtained. Allow to cool. Weigh 0·500 g of the finely ground, dried (110°C) sample on to a watch glass and brush the weighed portion of the sample on to the surface of the cold sodium hydroxide melt in the nickel crucible.

Gently tap the crucible to spread the sample evenly over the surface of the melt. Moisten the sample with ethanol (95%) and carefully evaporate the alcohol on a hot plate. Provided reasonable care is taken, this will prevent the light sample being blown from the crucible during the fusion.

Gently fuse over a mushroom burner, occasionally swishing the crucible, until the sample is dissolved and the melt is quiet, then increase the temperature to a dull red heat; 5 min is usually sufficient for the fusion.

Carefully cool the mass by running water round the outside of the crucible until the melt just solidifies. Place the still hot crucible in a 250-ml nickel beaker, and cover with a clock glass.

Raise the cover slightly, fill the crucible with water at the boiling point and quickly replace the cover glass. The crucible should be hot enough to boil the water and dissolve the fused mass. When the vigorous reaction is over, wash the cover glass and the sides of the beaker with hot water and remove the crucible with a pair of tongs, carefully rinsing it inside and out with hot water and scrubbing it with a "bobby".

Filter through a No. 541 Whatman paper, transferring the precipitate to the filter with a jet of hot sodium carbonate solution (10 g/litre). Wash thoroughly with hot sodium carbonate solution (10 g/litre).

Determination of Sulphate

To the filtrate from either of the above decompositions add about 5 ml of bromine water, and just acidify with hydrochloric acid ($d = 1·18$), added drop by drop, then add 2 ml in excess. Boil for about 2 min to expel carbon dioxide.

Add, drop by drop while boiling, 5 ml of barium chloride solution (100 g/litre), continue boiling for 2 min and allow to stand on a steam bath for 1 h. Allow to cool and stand overnight.

Filter through a No. 42 Whatman paper, transferring the precipitate to the filter with a jet of hot water and scrubbing the beaker with a "bobby". Wash free from chlorides with hot water. Discard the filtrate.

Transfer the precipitate and paper to a weighed platinum crucible, heat cautiously to dry the residue and char the paper, then burn off the carbon at a low temperature. Allow to cool.

Moisten the residue with a few drops of diluted sulphuric acid $(1 + 1)$ add about 5 ml of hydrofluoric acid $(40\% \text{ w/w})$ and evaporate slowly to dryness on a sand bath in a fume cupboard. Ignite the dry residue to 1000°C, cool and weigh.

Calculation—Weight of $BaSO_4 \times 0.3430 = $ weight of SO_3.

CHAPTER 19

DETERMINATION OF CARBON DIOXIDE (CARBONATE) IN CERAMIC MATERIALS

GENERAL

The determination of carbonate is not often undertaken in ceramic analysis except in the analysis of carbonate minerals, e.g. dolomite. Even in this instance the loss on ignition is normally made to suffice. Carbonate is often determined, not for its own sake but to allow for its effect on the determination of carbon content of clays etc.

PRINCIPLE OF THE METHOD

The carbonate is decomposed with hydrochloric acid and the carbon dioxide swept into previously weighed absorbant tubes by a stream of carbon dioxide-free air. The change in weight of the tubes gives the carbon dioxide content of the material.

REAGENTS

Additional Reagents

"Anhydrone" (magnesium perchlorate) or calcium chloride; anhydrous. "Carbosorb", or "Sofnolite"

APPARATUS REQUIRED

A. Moisture guard tube packed with "Anhydrone" or calcium chloride.
B. Carbon dioxide guard tube filled as in Fig. 22 (B), using "Carbosorb" or "Sofnolite" to absorb the carbon dioxide, and calcium chloride or "anhydrone" to absorb the released moisture.
C. Dropping funnel containing about 100 ml of hydrochloric acid $(1 + 9)$
D. 250-ml flask.
E. Condenser to minimize HCl spray passing over.
F. Bubbler containing sulphuric acid $(d = 1.84)$.
G. Empty bubbler to act as sulphuric acid guard tube.
H. and H'. "Carbosorb" U-tubes filled as in Fig. 22 (B).
J. "Anhydrone" guard tube.

PROCEDURE

Assemble the apparatus as shown in Fig. 22, suspending the U-tubes from a rigid bar by wires, and connecting each piece of glassware with short lengths of stout-walled rubber tubing, so that the ends of the glass tubing are in contact. Before the determination, the apparatus should be checked for gas leaks and blanks should be carried out.

Remove the two U-tubes, H and H', wipe them with a clean cloth and allow them to stand in the balance room for 30 min. Open the taps momentarily to the air and weigh the tubes individually. Weigh out 1·000 g of the

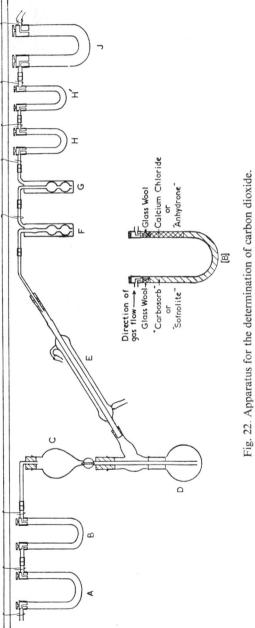

Fig. 22. Apparatus for the determination of carbon dioxide.

finely ground, dried (110°C) sample and transfer it to the flask D, which should be clean and dry. Replace tubes H and H′ in the absorption train, and open the taps of the U-tubes. Allow the acid in the dropping funnel C to run slowly into the flask D and gently heat the flask to boiling, regulating the speed of heating so that not more than one or two bubbles pass through the bubbler F each sec. It normally takes about 30 min to raise the temperature to boiling; boil for 30 min, allowing air to be drawn through the train at the above rate. Close the taps, remove tubes H and H′ and weigh, using the same procedure as before.

The increase in weight of tubes H and H′ gives the weight of carbon dioxide in the sample. The increase in weight of tube H′ should be almost negligible; if this is not so, the determination should be repeated, using a lower rate of gas flow, or after repacking the carbon dioxide absorption tubes H and H′.

CHAPTER 20

DETERMINATION OF SMALL AMOUNTS OF FLUORINE IN SILICATE MATERIALS

GENERAL

Two methods are described for the determination of small amounts of fluorine in such materials as Cornish stone. The first uses a pyrohydrolytic technique for the distillation of the fluorine, while the other relies on the chemical separation of interfering elements before the distillation. The former is the more elegant method and where this determination is likely to occur with even moderate frequency the cost of the apparatus is well justified.

A. Pyrohydrolytic Method

PRINCIPLE OF THE METHOD

The sample is mixed with a "catalyst", in this case vanadium pentoxide, and heated in a tube while passing steam. The fluoride is expelled as a mixture of HF and H_2SiF_6. This is condensed and the dilute acid solution adjusted in pH. The fluorine is then titrated with thorium nitrate solution using Alizarin red S as indicator.

REAGENTS

Additional Reagents

Alizarin red S
Chloracetic acid
Sodium fluoride
Starch: soluble
Thorium nitrate
Vanadium pentoxide

Prepared Reagent

Chloracetic acid buffer: Dissolve 22·7 g of chloracetic acid in 100 ml of water and titrate 50 ml of the solution with sodium hydroxide solution (250 g/litre), using phenolphthalein as indicator. Add the titrated solution to the remaining 50 ml and dilute to 1 litre.

Indicators

Alizarin red S
Phenolphthalein
Starch

Standard Solutions

Fluoride (1·0 mg F/ml)
Thorium nitrate (approx. 0·1N)

APPARATUS

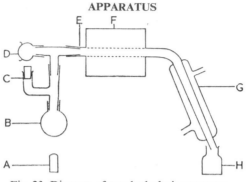

Fig. 23. Diagram of pyrohydrolysis apparatus.

A burner or electric mantle A is used to heat a 1- or 2-litre round-bottomed flask B to produce steam (Fig. 23). Plug C can be removed to allow the steam to dissipate to the atmosphere. Plug D can be removed to allow the introduction of the platinum boat into the silica tube E which is heated by the furnace F. The steam which has passed through the tube is condensed by the condenser G and the solution is collected in a 100-ml polythene bottle H with a mark at the 50-ml level.

The glass apparatus was obtained from Quickfit & Quartz Limited, and the silica tube from the Thermal Syndicate Limited. Catalogue details are given where applicable (Fig. 24).

PROCEDURE

The weight of material used for the determination should be varied with the nature of the sample and the V_2O_5 kept at more than about 50 times the anticipated weight of fluorine. Suggested weights are:

Cornish stones	0·2 g
Bodies	0·5 g
Feldspars	1·0 g
Lepidolites, etc.	0·1 g
(>2% F)	

Accurately weigh the appropriate amount of the finely ground, dried (110°C) sample and 0·2 g of vanadium pentoxide. Mix the sample and the vanadium pentoxide by grinding in an agate mortar, and transfer the mixture to a platinum boat.

Meanwhile heat the furnace to 800°C and boil the water in the flask allowing the steam to dissipate to the atmosphere.

Insert the platinum boat into the centre of the furnace and rapidly replace plugs D and C in that order. Increase the temperature of the furnace to 850°C while passing steam at a rate equivalent to 4–5 ml of condensate per min.

Collect 50 ml of the distillate in the polythene bottle and change the receiver. Repeat this collection; for most materials a total of 100 ml is normally sufficient.

Transfer each 50 ml distillate to a 500-ml conical flask marked to indicate a volume of 90 ml. Add 1 drop of phenolphthalein indicator, followed by

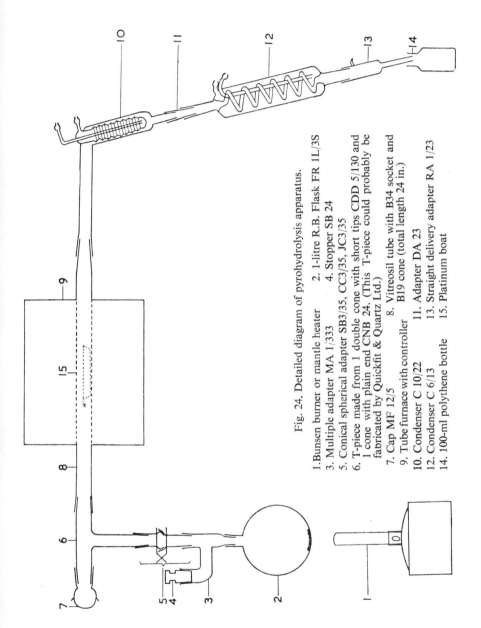

Fig. 24. Detailed diagram of pyrohydrolysis apparatus.

1. Bunsen burner or mantle heater   2. 1-litre R.B. Flask FR 1L/3S
3. Multiple adapter MA 1/333   4. Stopper SB 24
5. Conical spherical adapter SB3/35, CC3/35, JC3/35
6. T-piece made from 1 double cone with short tips CDD 5/130 and 1 cone with plain end CNB 24. (This T-piece could probably be fabricated by Quickfit & Quartz Ltd.)
7. Cap MF 12/5   8. Vitreosil tube with B34 socket and B19 cone (total length 24 in.)
9. Tube furnace with controller   11. Adapter DA 23
10. Condenser C 10/22   13. Straight delivery adapter RA 1/23
12. Condenser C 6/13   15. Platinum boat
14. 100-ml polythene bottle

sodium hydroxide solution (4 g/litre), drop by drop, until a permanent pink colour is obtained. Adjust the volume to 90 ml.

Add 10 ml of starch indicator, followed by diluted perchloric acid (1 + 9) drop by drop, until the colour is discharged, and add 1 drop in excess. Then add 0·5 ml of chloracetic acid buffer solution, followed by 5 drops of Alizarin red S indicator. Titrate with standard thorium nitrate solution (approx. 0·1N) until the first appearance of a faint pinkish tinge.

Calculate the fluorine content of the distillates, and from the total the fluorine content of the sample.

BLANK DETERMINATION

A blank determination should be carried out as above but omitting the sample. Normally half a drop (0·02 ml) of thorium nitrate solution per titration suffices.

B. Chemical Method

PRINCIPLE OF THE METHOD

The sample is fused in sodium carbonate and the melt extracted with hot water. The residue after filtering is discarded (for higher fluorine contents it may be desirable to repeat the fusion). Interfering elements are removed by a precipitation with zinc oxide and the fluorine distilled using sulphuric acid. After adjusting the acidity the fluorine is titrated with thorium nitrate using Alizarin red S as indicator.

REAGENTS

Additional Reagents

Alizarin red S
Chloracetic acid
Sodium fluoride
Starch: soluble
Thorium nitrate
Zinc oxide
Zinc sulphate

Prepared Reagents

Ammoniacol zinc carbonate: Add 25 g of zinc oxide and 50 g of ammonium carbonate solution to 60 ml of ammonia solution ($d = 0·88$). Shake occasionally for 30 min; then add slowly, with shaking, 110 ml of lukewarm water. Filter the solution when dissolution is complete.

Chloracetic acid buffer: Dissolve 22·7 g of chloracetic acid in 100 ml of water and titrate 50 ml of the solution with sodium hydroxide solution (250 g/litre), using phenolphthalein as indicator. Add the titrated solution to the remaining 50 ml and dilute to 1 litre.

Indicators

Alizarin red S
Methyl red
Phenolphthalein
Starch

Standard Solutions

Fluoride (1·0 mg F/ml)
Thorium nitrate (approx. 0·1N)

PROCEDURE

Preparation of the Solution for Distillation

Weigh 1·000 g of the finely ground, dried (110°C) sample into a platinum crucible, and mix with 3 g of anhydrous sodium carbonate. Cover the mix with a further 3 g of anhydrous sodium carbonate and transfer to a mushroom burner.

Raise the temperature gradually until frothing ceases, then complete the fusion at 1000°C for 15 min. Quench the melt by immersing the bottom half of the crucible in cold water and extract the melt in 100 ml of hot water contained in a 250-ml beaker.

Crush the cake with an agate pestle and digest on a steam bath until the melt has completely disintegrated. Filter through a No. 41 Whatman paper, transferring the residue to the filter with a jet of hot water and wash thoroughly with hot water. Discard the residue.

Evaporate the filtrate and washings to about 100 ml, then add, *with stirring*, to the boiling filtrate, 10 ml of zinc sulphate solution (180 g/litre). Continue boiling, *with stirring*, for 2–3 min. Cool to room temperature.

Filter through a No. 541 Whatman paper, wash once with cold water and reserve the filtrate and washings.

Wash the precipitate back into the precipitation beaker and add diluted sulphuric acid (1 + 1) until a solution just acid to methyl red is obtained and then add sodium carbonate, with stirring, until an excess of 1–2 g is present. Dilute to 50–60 ml and heat to boiling, *with constant stirring*.

Re-precipitate with zinc sulphate solution (180 g/litre) as before, cool, filter through a No. 541 Whatman paper and wash with a little cold water. Discard the residue.

Add diluted sulphuric acid (1 + 1) to the combined filtrates until the solution is slightly acid, and then add 2·5 ml of ammoniacal zinc carbonate solution. Boil the mixture down to about 100 ml and until no odour of ammonia is perceptible.

Cool and then filter through a No. 541 Whatman paper, wash once with cold water and reserve the filtrate. Dissolve the precipitate in the minimum amount of diluted sulphuric acid (1 + 1) and make up the volume of the solution to 100–150 ml with water.

Again add 2·5 ml of ammoniacal zinc carbonate solution and boil down to about 70–80 ml and until no odour of ammonia is perceptible. Cool, filter and wash with a little cold water. The combined filtrates are now ready for distillation.

Distillation Apparatus (Fig. 25)

A 250-ml Claisen flask (preferably Pyrex) with the bent neck about 25mm dia. is clamped in a retort stand and rests on a piece of asbestos sheeting about 150mm square with a central hole of 50mm dia.

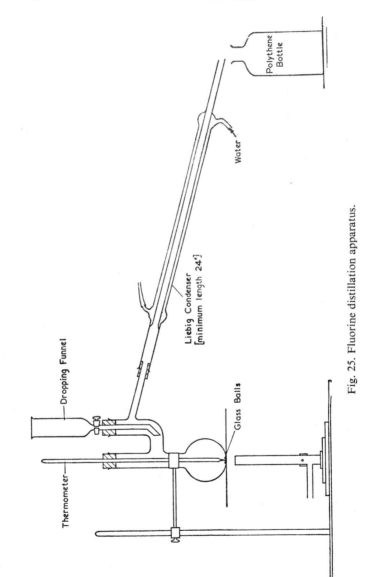

Fig. 25. Fluorine distillation apparatus.

The bent neck of the flask is closed with a rubber bung through which passes the stem of a 100-ml dropping funnel, and a thermometer (range at least 0–150°C) is inserted through a rubber bung in the straight neck so that its bulb is near the bottom of the flask. A Liebig condenser (total length at least 600 mm with a water-cooled length of at least 350 mm) is attached to the side arm of the flask by another rubber bung.

Prior to distillation, place about six glass balls of approximately 5 mm dia. in the flask to prevent bumping. Collect the distillate in a 250- or 350-ml polythene bottle with marks on the side to indicate volumes of 20 ml and 200 ml.

Distillation of Fluorine

Transfer the distillate to the Claisen flask, fit the thermometer and the dropping funnel and connect the flask to the condenser. Turn on the condenser cooling water. Loosen any clamps holding the flask and condenser and, while swirling the liquid in the flask, add 35 ml of sulphuric acid ($d = 1.84$) from the dropping funnel. Rinse the funnel with 1–2 ml of water and close the tap. Tighten the clamps in the apparatus and heat the flask.

Distil until at least 20 ml of distillate has collected and the liquid is boiling at 120°C or over. Change the receiver.

Return the distillate to the dropping funnel but not to the flask (this return of the distillate is to eliminate contamination from the spray made during the first addition of the acid).

Continue the distillation and when the temperature of the liquid reaches 135°C add the contents of the dropping funnel, drop by drop (followed by water), to maintain a temperature of 135 ± 5°C. Collect 200 ml of distillate and store until required, closing the bottle with a bung.

When the distillation flask is reasonably cool, wash it once or twice with water, then with hot, strong sodium hydroxide solution and again wash well with distilled water. This is to remove any gelatinous silica that may have been deposited on the flask during the distillation and which would tend to retain fluorine in subsequent distillations.

Titration of the Distillate

Transfer the distillate to a 500-ml conical flask marked to indicate a volume of 90 ml. Add 1 drop of phenolphthalein indicator followed by sodium hydroxide solution (4 g/litre), drop by drop, until a permanent pink colour is obtained.*

Evaporate until the volume is about 90 ml. If the distillate has become bleached during evaporation or dilution, add one more drop of phenolphtha-

---

*Add not more than 5 ml of sodium hydroxide solution (4 g/litre). If more is required continue with sodium hydroxide solution (4 g/litre) then dilute to 250 ml in a volumetric flask and take an aliquot equivalent to about 5 ml of sodium hydroxide solution (4 g/litre). If the fluorine content is expected to be greater than 1·0–1·5 % it is preferable to take a smaller sample calculated to contain about 10 mg F.

I*

lein indicator and enough sodium hydroxide solution (40 g/litre) to give a permanent pink colour.

Add 10 ml of starch indicator followed by diluted perchloric acid (1 + 9), drop by drop, until the colour is discharged. Add 1 drop in excess and then add 0·5 ml of chloracetic acid buffer solution. Add 5 drops of Alizarin red S indicator and titrate with standard thorium nitrate solution (approx. 0·1N) until the first appearance of a faint pinkish tint.

Calculate the fluorine content of the solution, and from the total the fluorine content of the sample.

CHAPTER 21

DETERMINATION OF SMALL AMOUNTS OF PHOSPHATE IN CLAYS Etc.

PRINCIPLE OF THE METHOD

The silica in the sample is removed by treatment with hydrofluoric acid, the residue fused in sodium carbonate and the melt dissolved in hydrochloric acid. Phospho-molybdic acid is then formed by the addition of sodium molybdate and this is precipitated as the quinoline salt. The precipitate is filtered off, dissolved in an excess of standard alkali and the excess back-titrated with hydrochloric acid.

REAGENTS

Additional Reagents

Quinoline: boiling range 235–239°C
Sodium molybdate
Thymol blue

Prepared Reagent

Quinoline (2% v/v): Add 20 ml of quinoline (boiling range 235–239°C) to about 800 ml of hot water acidified with 25 ml of hydrochloric acid (d = 1·18), stirring constantly if cloudy. Cool, add some paper pulp and stir vigorously. Allow to settle and filter under suction through a paper pulp pad to remove traces of oily matter, but do not wash. Dilute the filtrate to 1 litre.

Indicator

Phenolphthalein-thymol blue mixed indicator

Standard Solutions

Hydrochloric acid (approx. 0·5N)
Sodium hydroxide (approx 1N).

PROCEDURE

Decomposition of the Sample

Weigh 1·000 g of the finely ground, dried (110°C) sample into a platinum crucible, add 5 ml of diluted sulphuric acid (1 + 9) followed by 10 ml of hydrofluoric acid (40% w/w) and evaporate slowly to dryness on a sand bath, in a fume cupboard, taking care to avoid spurting.

Allow the crucible to cool, add a further 5 ml of diluted sulphuric acid (1 + 9) and 10 ml of hydrofluoric acid (40% w/w) and evaporate to dryness. Finally add a further 5 ml of diluted sulphuric acid (1 + 9) and evaporate to dryness to remove the last traces of fluoride.

Mix the residue with 4 g of anhydrous sodium carbonate and transfer the crucible to a mushroom burner and cover with a lid. Slowly raise the temperature of the crucible and contents until a quiet fusion is obtained. Fuse for 15 min at 1000°C and allow to cool.

Dissolve the melt in 100 ml of water to which 27 ml of hydrochloric acid ($d = 1 \cdot 18$) has been added; this gives an excess of about 20 ml.

Determination of Phosphate

Transfer the solution to a 500-ml conical flask and adjust the volume to about 150–200 ml. Add a small quantity of paper pulp, followed by 30 ml of sodium molybdate solution (150 g/litre), stirring during the addition. Heat to boiling, add 2–3 drops of quinoline solution (2% v/v) from a burette, boil again and add drop by drop, 2 ml of quinoline solution (2% v/v). Continue boiling and add quinoline solution (2% v/v) in 2 ml increments, heating to boiling about every 15 ml until a total of 60 ml has been added.

Transfer the flask to a steam bath for 45 min and then cool in running water to less than 20°C.

Filter under suction through a No. 42 Whatman paper in a Buchner funnel, wash twice by decantation with 10-ml portions of diluted hydrochloric acid $(1 + 9)$, and then with cold water until the washings give an alkaline reaction with 1 drop of sodium hydroxide solution (approx. 1N) and a few drops of phenolphthalein–thymol blue mixed indicator. Discard the filtrate and washings.

Transfer the precipitate and paper back to the precipitation flask and dissolve in 25 ml of standard sodium hydroxide solution (approx 1N). Dilute to about 100 ml, add 6 drops of phenolphthalein–thymol blue mixed indicator and titrate with standard hydrochloric acid (approx. 0·5N) from purple through green to yellow.

Calculation—Convert the volume of sodium hydroxide used into its equivalent volume of 0·5N solution.

With a 1·0 g sample, 1 ml 0·5N NaOH $\equiv 0 \cdot 1366\%$ P_2O_5.

CHAPTER 22

DETERMINATION OF FERROUS OXIDE IN
HIGH-SILICA MATERIALS AND ALUMINOSILICATES

GENERAL

A knowledge of the ferrous oxide content of raw materials and refractories is sometimes of industrial importance. The method described in this Chapter is commonly used for comparative purposes, although the absolute accuracy is in doubt. Results will often tend to be low because of the failure to prevent oxidation of ferrous iron during grinding, storage and decomposition of the sample. Little care is taken in this method except in restricting access of air to prevent the latter, other methods screen the solution with carbon dioxide or add an oxidant such as vanadium pentoxide. This latter technique oxidizes the ferrous iron in situ and the excess oxidant is determined. High results can sometimes be obtained by the presence of carbonaceous matter which reduces the titrant.

It will be seen that no wet method can be completely satisfactory, but a full review of methods may be found in John A. Maxwell (1968) "Rock and Mineral Analysis". Interscience.

PRINCIPLE OF THE METHOD

The sample is decomposed in hydrofluoric acid in the absence of air and the excess hydrofluoric acid complexed with boric acid. The ferrous iron in the sample is then titrated with potassium dichromate.

REAGENTS

Additional Reagent

Barium diphenylamine sulphonate.

Prepared Reagent

Sulphuric acid-phosphoric acid mixture: Cautiously add 75 ml of sulphuric acid ($d = 1\cdot84$) and 75 ml of phosphoric acid ($d = 1\cdot75$) to 350 ml of water, cooling the solution and keeping it well mixed during the addition of the acids.

Indicator

Barium diphenylamine sulphonate

Standard Solution

Potassium dichromate ($0\cdot1$N).

PREPARATION OF SAMPLE

The sample should be ground to pass completely through a 60-mesh B.S. test sieve. Too fine grinding may promote oxidation of the iron.

PROCEDURE

Determination of Hygroscopic Water

Heat 5–10 g of the prepared, air-dried sample in an air oven at 110°C to constant weight. The sample should be spread in a thin layer. Drying for 3–4 h is usually sufficient. The determination of hygroscopic water should be commenced at the same time as the samples are weighed for the determination of ferrous oxide.

Determination of Ferrous Oxide

Weigh 1·000 g of the prepared, air-dried sample into a platinum crucible of about 30 ml capacity, add 10 ml of diluted sulphuric acid (1 + 1) and stir with a glass rod. If the sample is not stirred it may cake on the bottom of the crucible and escape the attack of the acid.

Add about 10 ml of hydrofluoric acid (40% w/w) and cover the crucible with a tight-fitting lid. Place the crucible on a sand bath in a fume cupboard in such a position that the contents boil gently. Extreme care is required here to avoid boiling the acid over the top of the crucible. About 10 min is required for the decomposition of the sample.

Remove the crucible, with the lid still in position, from the sand bath and plunge it into a 600-ml beaker containing 300 ml of cold, freshly-boiled distilled water in which 15 g of boric acid has been dissolved. (On cooling, the excess boric acid will be precipitated). Swirl the beaker gently to remove the lid from the crucible and mix its contents with those of the beaker.

Add 15 ml of sulphuric acid–phosphoric acid mixture and 10 drops of barium diphenylamine sulphonate indicator. Titrate to a purple colour with standard potassium dichromate solution (0·1N).

Calculation—1 ml 0·1N $K_2Cr_2O_7$ ≡ 0·00718 g FeO.

CHAPTER 23

ANALYSIS OF MATERIALS CONTAINING SIGNIFICANT AMOUNTS OF FLUORINE AND/OR PHOSPHATE, e.g. CORNISH STONE

Cornish stones as used in the British pottery industry can contain up to $1 \cdot 5\%$ of fluorine and $1 \cdot 0\%$ of phosphate (P_2O_5). There is also an increasing tendency to use various phosphates as bonding agents in the manufacture of refractories. The presence of these interfering constituents needs to be considered when choosing suitable methods of analysis, since, if some of the methods described are used without modification inaccuracies will result, though, for most industrial purposes it has been possible to ignore them.

In the classical method the total usually fell within the permissible limits, owing to negative errors in one part of the analysis being counter-balanced by positive errors elsewhere. With recent methods tending to be more specific, errors are often negative and analytical totals are less likely to be satisfactory. The effect of the presence of these constituents on the determination of other oxides for the various methods likely to be used is considered below.

The determination of fluorine or phosphorus at these relatively low levels does not usually present a problem and is dealt with in other Chapters.

CLASSICAL METHOD

Fluorine

Some, possibly all, of the fluorine will normally be lost during the ignition to determine "loss on ignition", when silicon tetrafluoride will be evolved. Experience has shown that this can occur at different temperatures with different types of Cornish stone and that the reaction does not always proceed to completion so that it is not always possible to calculate the loss of silica or an adjusted "loss on ignition" figure. However, if the ignition is sufficiently prolonged until the loss of weight is constant it is probably correct to assume that all the fluorine has been volatilized, hence the loss on ignition value will be too high by the weight of silicon tetrafluoride evolved. Now if the ignited sample is used for the main analysis (SiO_2, R_2O_3, etc), the determined silica content will be low by the amount of silica lost as silicon tetrafluoride. Even if a fresh sample is used, much of the fluorine may be lost during the successive evaporations from the acid solutions during the determination of silica.

Phosphorus

The phosphorus pentoxide will be precipitated with the ammonia group oxides, and unless it is separately determined and allowed for, the alumina content as calculated will be higher by the weight of phosphorus pentoxide present. If the determination is carried out directly by the oxine gravimetric procedure, the alumina figure will be correct, since phosphate does not interfere with this determination. Thus the presence of phosphate does not present a very great problem if this course is adopted (Chapter 10). The determination of phosphate content may be carried out using the procedure described in Chapter 21.

SINGLE DEHYDRATION METHOD

Fluorine

The same arguments apply to this method as were applied to the classical determination. Some or all of the fluorine will be lost during the loss on ignition, resulting in difficulty in achieving reliable analytical totals and the accurate determination of silica. Again, the use of a separate sample does not solve the problem as silicon tetrafluoride may be lost during the evaporation to dryness for the determination of silica.

Phosphorus

As the determination of alumina is now carried out directly, either by an oxine gravimetric finish or by the EDTA volumetric finish, and as neither of these methods is affected by the presence of phosphate, the phosphate content of Cornish stones no longer constitutes a problem in the determination of alumina. However, the silica determination includes a determination of residual silica remaining in solution after a single dehydration. This determination is carried out colorimetrically as the yellow silicomolybdate and it is here that phosphate interferes by the formation of the yellow coloured phosphomolybdate, thus resulting in apparently high values for the silica contents. This difficulty can be overcome by using the molybdenum-blue procedure.

COAGULATION METHOD

Fluorine

Fluorine is much less likely to be lost by this method as, on making acid, little or no evaporation occurs. The incorporation of boric acid in the flux will tend to complex the fluorine as fluoboric acid, in which form interference with the alumina determination is minimized. However, the use of the gravimetric oxine method for the determination of alumina may be preferable since fluoride interferes in the EDTA procedure.

It is unlikely that any significant error will occur in the determination of silica provided that an unignited sample is fused directly, as no silicofluoride will be lost by evaporation of the acid solution, and as the fluoride will be present mainly complexed as fluoboric acid.

Phosphorus

The same remarks apply as for the single dehydration method.

RAPID METHOD

Fluorine

Fluorine does not interfere with either the determination of silica by the quinoline silicomolybdate method or alumina by the oxine method. It is possible that there may be some small amount of interference with the determination of titania and ferric oxide.

Phosphorus

Phosphorus will interfere with the determination of silica as quinoline silicomolybdate by the precipitation of the analogous quinoline phosphomolybdate. The results can of course be corrected easily by subtracting from the weight of precipitate, the weight equivalent to quinoline phosphomolybdate, calculated from the content of phosphate determined separately.

The factor for converting the weight of P_2O_5 to the equivalent weight of quinoline phosphomolybdate, $(C_9H_7N)_6.P_2O_5.24\ MoO_3.3H_2O$, is 31·18.

SPECTROPHOTOMETRIC METHOD

Fluorine

Fluorine does not interfere with the determination of silica as yellow silicomolybdate, but will interfere with the determination of alumina with Solochrome cyanine, reducing the intensity of the colour of the complex severely. The error is liable to be so large that it renders the method of very doubtful value for samples containing as much fluorine as is often found in Cornish stone.

Phosphorus

Phosphorus interferes with the determination of silica as described above by forming a coloured phosphomolybdate and thereby producing a more intensely coloured solution than would otherwise be the case. Again it is possible to correct for this interference, but in view of the inherent lack of precision in the spectrophotometric method it is very doubtful if any great reliance can be placed on the final figures.

BERZELIUS METHOD

This method of analysis of fluorine-bearing materials was developed many years ago and is still required as a proven procedure to be adopted when other possible simpler approaches fail. It has the advantage that it is theoretically sound and that the determinations of silica and fluorine are made directly whereas none of the other methods mentioned above provides a direct determination of silica. On the other hand the method is very long, tedious and cumbersome. Even when the results are obtained they are of very doubtful accuracy except perhaps in the hands of a thoroughly experienced and competent analyst. It is very difficult to estimate the potential errors of the method and consequently to state an anticipated degree of accuracy for the results.

The technique is described in detail in many standard text books including Mellor and Hillebrand (see page xi).

CONCLUSIONS

The coagulation method provides a useful method for the analysis of materials containing small amounts (less than 2 %) of fluorine and phosphate. Errors will be small, particularly if the oxine gravimetric determination of alumina is used after the cupferron/chloroform separation.

The classical method and the single dehydration method are both unsound when applied to this type of material, since the fluorine content renders an accurate determination of the silica content almost impossible. Errors due to the presence of phosphate can be overcome by allowing for this constituent in the calculation of the silica or alumina contents.

The spectrophotometric method is not worth applying to Cornish stones as its already low reproducibility and accuracy are further reduced by the presence of either fluorine or phosphate.

The rapid method, although theoretically less accurate than most of the other possible methods, is in this instance probably the most convenient and accurate. The results are not liable to interference by fluorine in any way and the phosphate content of the sample, although interfering with the determination of the silica content, is easily determined and the correct silica content found by a simple calculation.

A determination of loss on ignition can naturally be carried out without any difficulty, but it is almost impossible to interpret the meaning of the result and to fit it into the final analysis. The determination is empirical in the sense that a temperature of 1000°C is chosen as an arbitrary figure, a temperature at which all water of composition and carbonaceous matter can be expected to be removed. In the case of Cornish stone, however, although water and carbonaceous matter have been removed, fluorine is also lost, and it is impossible to know if all the fluorine has been removed. It is thus impossible to use the figure obtained for calculation of the rational analysis or to allow for the silica lost during the ignition. Our experience has been that Cornish stones fall into at least three categories: (1) those which appear to lose almost all the water etc. at 600°C: (2) those losing all the water etc. at 800°C: and (3) those which do not lose all their water until they are heated to a temperature higher than 800°C. With the first group it is possible to carry out a loss on ignition at 600°C with little danger of losing fluorine, with the second group it is likely that the loss of fluorine at 800°C is tolerable, but with the third group it is almost impossible to obtain a loss on ignition figure that has any significance. It is idle to pretend that this state of affairs is satisfactory, but the only final solution would appear to be analysis of the gases evolved in the course of the ignition. This can be done, but there has not been sufficient interest in the analysis of Cornish stone to a great level of accuracy to justify this action. It should be noted that the stone is added to a body which is then fired to a temperature considerably in excess of 1000°C so that the ultimate interest of the ceramist lies in the analysis of the stone in its ignited state, although it is obviously helpful to have a reasonable idea of its composition before firing.

CHAPTER 24

DETERMINATION AND ANALYSIS OF SOLUBLE SALTS IN CLAYS AND BRICKS

The determination of the amount of soluble salts in clays and fired products is essentially an empirical test. The results will clearly depend on the fineness of the material, the ratio of water to solid and the temperature at which the extraction is carried out; in addition, there are probably a number of other factors which are not so obvious. It is thus necessary to specify the details of the test, and to carry out the test paying attention to these details.

At present there are two different specifications for the test, one a British Standard for fired products, in particular building bricks, and the other for clays. These two specifications differ in almost all respects and will need to be described separately, but the method of analysing the salts is similar.

PRINCIPLE OF THE METHOD

The principle of the B.Ceram.R.A. method for the determination of soluble salts in clay is merely to suspend the clay for a longer time period (24 h) at a high rate of dilution.

The British Standard method (B.S. 3921: 1965) on the other hand, in the determination of soluble salts in fired bricks prefers to increase the concentration and shorten the time period. This is to minimize the solubility of calcium sulphate, the amount of which is estimated by carrying out a separate acid attack in order to determine soluble sulphate.

It is now customary to determine only soluble alkali, lime, magnesia and sulphate contents but clearly other determinations may be carried out, if required, by appropriate methods.

REAGENTS

Prepared Reagents

Ammonium nitrate (10 g/litre): Dilute 10 ml of nitric acid ($d = 1.42$) to about 200 ml. Add diluted ammonia solution ($1 + 1$) until the solution is faintly alkaline to methyl red. Dilute to 1 litre.

Silver nitrate (20 g/litre): Dissolve 5 g of silver nitrate in 250 ml of water. Store in a dark bottle.

Indicators

Bromophenol blue
Calcein (screened)
Methylthymol blue complexone

Standard Solution

EDTA (5 g/litre)

BRITISH STANDARD METHOD (B.S. 3921: 1965)
(FOR FIRED BRICKS)

Preparation of the Sample

The powdered sample required for analysis is obtained by drilling holes in ten bricks with a masonry drill not larger than $\frac{1}{4}$ in. dia. The holes should be approximately evenly spaced over the bed face of each brick and carried to a depth approximately equal to half the depth of the brick. The number of holes should be sufficient to give a sample of approximately 25 g of powder passing through a 100-mesh B.S. test sieve. Material from the drillings which does not pass the sieve immediately should be ground in a suitable mortar until the whole sample passes through. A magnet should be used to remove any iron which may have contaminated the sample during drilling.

Alternatively, fragments representing the interior and the exterior of the bricks amounting to at least one tenth of each brick should be crushed to produce about 5000 g of material passing a sieve with an aperture not greater than B.S. No. 5. This should be mixed and then reduced by coning and quartering to about 300 g which should be then all ground to pass a sieve not greater than B.S. No. 22. The latter sample should again be reduced by coning and quartering and ground to pass a 100-mesh B.S. test sieve.

A magnet should be used to remove any iron that may have contaminated the sample during crushing. The sample should be dried at 110°C.

Determination of Acid-soluble Sulphate

Weigh 2 g of the sample and transfer to a 250-ml beaker and cover with a clock glass. Through the lip of the beaker introduce 150 ml of hydrochloric acid (1 + 9) and heat to boiling, add half a Whatman ashless tablet or equivalent and boil for 10 min, stirring to prevent bumping.

Cool, filter through a sintered-glass Buchner funnel and wash five or six times with hot distilled water.

Add 1 or 2 drops of methyl red indicator and diluted ammonia solution (1 + 1) dropwise until just neutral then add immediately 25 drops of hydrochloric acid ($d = 1\cdot18$) followed by 3 ml of bromine water (saturated). Heat to boiling, boil for 2 min and, while boiling, slowly add from a pipette 10 ml of barium chloride solution (100 g/litre).

Continue boiling for about 2 min, transfer to a steam bath for 1 h and allow to cool.

Stand overnight and filter through a No. 42 Whatman paper (or equivalent). Wash with hot water until free from chlorides.

Transfer the precipitate and paper to a weighed platinum crucible, heat gently to dry and char the paper, and finally ignite to 1000°C for 30 min, cool and weigh.

$$\text{Weight of BaSO}_4 \times 0\cdot4115 = \text{Weight of SO}_4^{--}$$

NOTE: The acid-soluble sulphate may be assumed to correspond fairly closely to the total quantity of sulphate which could be obtained from the brick sample on long continued extraction with water. This quantity is therefore relevant to the assessment of the liability of the brick material to cause sulphate expansion in Portland Cement.

Extraction of Water-soluble Salts

The extraction of soluble salts should be carried out at room temperature.

Weigh 10 ± 0.05 g of the sample and transfer it to a 150-ml polythene bottle. Add 100 ml of cold distilled water, close the bottle with a screw-type polythene top and stir the contents for 60 min with a magnetic stirrer using a polythene-covered follower (alternatively a rotary shaker at about 30 revolutions per min may be used).

The suspended sample is filtered and the filtrate collected in a clean dry flask without washing the residue. The filtering medium is used dry; the following alternatives are permissible.

(i)   Sintered-glass Buchner funnel (grade 4)
(ii) Centrifuge
(iii) Filter candle
(iv) Filter paper.

The filtrate must be clear.

Determination of radicals

Any recognized procedure may be used for the determination of Ca, Mg, Na, K—reported as the element. The following method of analysis is recommended but is not mandatory. Results should be reported to the nearest 0.01%.

Calcium: Pipette a 10-ml aliquot of the soluble salt extract into a 500-ml conical flask. Add 20 drops of hydrochloric acid ($d = 1.18$) followed by 10 ml of potassium hydroxide solution (250 g/litre) and dilute to about 200 ml. Add about 0.015 g of Calcein indicator. Titrate with standard EDTA solution (5 g/litre) from a 10-ml semi-micro burette, the colour change being from fluorescent green to pink.

Magnesium: Pipette a 10-ml aliquot of the soluble salt extract into a 500-ml conical flask. Add 20 drops of hydrochloric acid ($d = 1.18$) followed by 10 ml of ammonia solution ($d = 0.88$) and dilute to about 200 ml with water. Add about 0.05 g of methylthymol blue complexone indicator. Titrate with standard EDTA solution (5 g/litre) from a semi-micro burette, the change being from blue to colourless.

The volume of EDTA used for the titration of lime is subtracted from the volume of EDTA used for this titration. The remainder represents the volume of EDTA required for the titration of magnesium.

Sodium and Potassium: In view of the large amount of lime which is likely to be present in the soluble salt extract it is advised that alkalis should be determined by a method similar to that given for aluminosilicate materials, i.e. by the addition of caesium and aluminium sulphates. The method is given in Chapter 16 under "Soluble Salts".

BRITISH CERAMIC RESEARCH ASSOCIATION METHOD
(FOR RAW MATERIALS)

Preparation of the Sample

The sample prepared for analysis should be ground to pass completely through a 120-mesh B.S. test sieve. A non-metallic (e.g. nylon bolting-cloth) sieve is preferable.

Extraction of the Soluble Salts

Weigh 50 g of the undried sample and transfer to a 1-litre Stohmann shaking flask. Add 1 litre of water, close the neck of the flask with a rubber bung and shake for 24 h. A rotary shaker is suitable. Filter off the suspended clay through a pulp paper pad, or porcelain filter candle under suction (Fig. 26). Collect the filtrate in a clean dry Buchner flask. *Do not wash.*

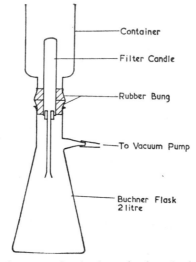

Fig. 26. Apparatus for the determination of soluble salts.

Determination of the Total Soluble Salts

Transfer a 200-ml aliquot of the solution from the extraction to a 250-ml beaker, boil down to about 40–50 ml and transfer to a weighed platinum basin of about 75-ml capacity, washing the beaker with a jet of hot water.

Evaporate the solution to dryness, place the basin and residue in an air oven at 110°C for 1 h, cool and weigh. Record the difference in weight as "total soluble salts".

Determination of Loss on Ignition (water of crystallization, carbonaceous matter, etc.)

Ignite the basin after the above weighing to about 550°C for 30 min, cool and weigh. Record the difference between the weight of the basin plus total soluble salts and this weight as the loss on ignition.

Determination of Silica (when required)

Add 25 ml of diluted hydrochloric acid (1 + 9) to the residue and evaporate to dryness on a steam bath. Cover the basin with a clock glass and bake in an air oven at 110°C for 30 min.

Allow to cool, then drench the residue with about 5 ml of hydrochloric acid ($d = 1 \cdot 18$). Allow to stand for a few minutes, add about 25 ml of hot water and digest on a steam bath for 5 min to dissolve the salts.

Filter through a No. 42 Whatman paper, transferring the residue to the paper with a jet of hot water and scrubbing the basin with a "bobby". Wash free from chlorides with hot water. Reserve the filtrate and washings for the determination of the ammonia group oxides.

Transfer the residue and paper to a weighed platinum crucible, heat gently to dry the residue and char the paper. Burn off the carbon at a low temperature and finally ignite at 1200°C for 10 min. Cool and weigh.

Moisten the weighed residue with a few drops of diluted sulphuric acid (1 + 1) and add about 5 ml of hydrofluoric acid (40% w/w). Evaporate to dryness on a sand bath in a fume cupboard. Ignite the dry residue at the full heat of a Méker burner for about 5 min. Allow the crucible to cool, and weigh. Subtract the weight of the crucible plus residue from the weight of the crucible plus silica and residue, to obtain the weight of the pure silica. Reserve the crucible and residue.

Determination of Ammonia Group Oxides

Evaporate the filtrate reserved from the determination of silica to a volume of 100–150 ml. Add ammonia solution ($d = 0 \cdot 88$), drop by drop, until the solution is just alkaline to bromophenol blue and boil for 2 min.

Filter through a No. 40 Whatman paper, transferring the precipitate to the paper with a jet of hot, faintly ammoniacal ammonium nitrate solution (10 g/litre) and scrubbing the beaker with a "bobby". Wash the precipitate free from chlorides with hot, faintly ammoniacal ammonium nitrate solution (10 g/litre). Discard the filtrate and washings.

Transfer the precipitate and paper to the platinum crucible reserved from the determination of silica, heat gently to dry the residue and char the paper. Burn off the carbon at a low temperature and finally ignite at 1200°C for 10 min. Cool and weigh.

Subtract the weight of the empty crucible from this weight to obtain the weight of the total ammonia group oxides.

Determination of Lime

Transfer a 50-ml aliquot of the extraction solution to a 500-ml conical flask, and add a few drops of hydrochloric acid ($d = 1 \cdot 18$) followed by 2 ml of diluted triethanolamine (1 + 1). Unless iron has been noted in the solution as a post precipitate of ferric hydroxide, the addition of both the acid and the triethanolamine can be omitted.

Add 10 ml of potassium hydroxide solution (250 g/litre) and dilute to about 200 ml. Add about 0·015 g of screened Calcein indicator and titrate with standard EDTA solution (5 g/litre) from a semi-micro burette, the colour change being from fluorescent green to pink.

Determination of the Sum of Lime and Magnesia

Transfer a 50-ml aliquot of the extraction solution to a 500-ml conical flask and add a few drops of hydrochloric acid ($d = 1.18$) followed by 2 ml of diluted triethanolamine (1 + 1). Unless iron has been noted in the solution, as a post precipitate of ferric hydroxide, the addition of the triethanolamine can be omitted.

Add 10 ml of ammonia solution ($d = 0.88$) and dilute to about 200 ml. Add about 0.04 g of methylthymol blue complexone indicator and titrate with standard EDTA solution (5 g/litre) from a semi-micro burette, the colour change being from blue to colourless.

Calculation of Magnesia

The volume of EDTA used for the titration of lime is subtracted from the volume of EDTA used for the titration of the sum of lime and magnesia. The remainder represents the volume of EDTA used for the titration of magnesia.

Determination of Alkalis

Determine the alkali content of the solution flame-photometrically by the method described in Chapter 16 under "Soluble Salts".

Determination of Sulphate

Transfer a 200-ml aliquot of the extraction solution to a 400-ml beaker, add about 5 ml of bromine water and acidify with hydrochloric acid ($d = 1.18$), drop by drop, adding about 20 drops in excess.

Heat to boiling and boil for 2 min to expel carbon dioxide. While boiling, add slowly about 5 ml of barium chloride solution (100 g/litre) and continue boiling for 2 min. Then transfer to a steam bath and keep hot for 1 h. Cool and stand overnight.

Filter through a No. 42 Whatman paper, transferring the precipitate to the filter with a jet of hot water and scrubbing the beaker with a "bobby". Wash with hot water until free from chlorides. Discard the filtrate and washings.

Transfer the precipitate and paper to a weighed platinum crucible, heat gently to dry the precipitate and char the paper. Burn off the carbon at a low temperature and finally ignite at 1000°C for 30 min. Cool and weigh.

Calculation—Weight of $BaSO_4 \times 0.3430$ = weight of SO_3.

Determination of Chloride

Carry out this determination in subdued light.

Transfer a 200-ml aliquot of the extraction solution to a 400-ml beaker and just acidify with nitric acid ($d = 1.42$) to the change point of bromophenol blue and add 2 ml in excess. Add 5 ml of silver nitrate solution (20 g/litre) and heat to boiling. Boil for 5 min to coagulate the precipitate, and check that precipitation is complete by adding a further few drops of silver nitrate solution (20 g/litre).

Stand for at least 4 h or preferably overnight, in the dark. Filter through a weighed crucible with a sintered-glass mat (porosity 4) and wash thoroughly with cold, diluted nitric acid (1 + 99).

Dry in an air oven at 110°C for 2 h, cool and weigh.

Calculation—Weight of AgCl × 0·2474 = weight of Cl

EXPRESSION OF RESULTS

Results are usually expressed as percentages.

CHAPTER 25

DETERMINATION OF RESISTANCE TO ACIDS, AND ACID-SOLUBLE IRON

Both these tests are carried out according to B.S. 784: 1953, "Methods of Test for Chemical Stoneware".

The specification for acid resistance test is far from satisfactory, allowing a wide divergence of interpretation. It is difficult therefore to correlate results between laboratories. So far no satisfactory alternative has been suggested. Thus, if concordant results are to be obtained, it is essential within the individual laboratory to control the test more rigidly than is specified.

The test for acid soluble iron is somewhat better, but the technique used is rather crude and should be capable of improvement; this, however, is a matter for the appropriate British Standards Committee.

DETERMINATION OF RESISTANCE TO ACIDS

Preparation of Sample

The sample is taken from the body of the material and should be free from glaze.

Crush the sample in a stoneware mortar, sieve through an 18-mesh and a 25-mesh B.S. test sieve. The portion to be used for the test is that which passes the former and remains on the latter. At least 30 g of sample should be collected and washed free from dust. Place the material in a porcelain basin and add about 150 ml of water for each 30 g of sample. Heat the basin and contents on a sand bath until the water is at boiling point. Continue to heat for 1 h, taking care to avoid spurting. Decant the water and rinse the particles four times with cold water. Dry the residue at 110°C to constant weight; 4 h is normally sufficient.

Procedure

Weigh between 24·9 and 25·1 g of the prepared sample to an accuracy of 1 mg and place in a 110-mm porcelain basin. Add the following mixture:

7 ml of nitric acid ($d = 1·42$)
13 ml of sulphuric acid ($d = 1·84$)
65 ml of water.

Transfer the basin and contents to a sand bath in a fume cupboard and heat carefully to prevent loss by spurting until all the nitric acid has been evaporated and the sulphuric acid begins to fume strongly. The process should be arranged to take about 2 h.

Cool the basin and contents and add:

90 ml of water
10 ml of nitric acid ($d = 1·42$).

Repeat the heating process until the sulphuric acid again begins to fume strongly. Cool the basin and contents and then carefully decant the acid.

Add about 150 ml of water and heat to the boiling point. The cycle of decantation, addition of fresh water, boiling and decantation is repeated until the decanted liquor shows no trace of sulphate when tested with barium chloride solution (100 g/litre). Care should be taken to see that no particles are lost during these washings, etc. Dry the sample to constant weight at 110°C.

Calculation—Resistance to acids $= \dfrac{\text{Final weight in grams}}{\text{Initial weight in grams}} \times 100$

DETERMINATION OF ACID-SOLUBLE IRON

Reagents

Prepared Reagents

Ammonium acetate (approx. 10%): Dilute 140 ml of acetic acid (glacial) to 2000 ml with water and add carefully 140 ml of ammonia solution ($d = 0.88$). Mix, cool and adjust to approximately pH 6 either with acetic acid or ammonia solution.

1:10-phenanthroline (10 g/litre): Prepare enough solution for immediate use at a concentration of 0·1 g of 1:10 phenanthroline hydrate in 10 ml of diluted acetic acid (1 + 1).

Preparation of Sample

Break down the sample, free from glaze, in a stoneware mortar to pass a 10-mesh B.S. test sieve. Quarter the sample and grind an appropriate amount in an agate mortar. Collect for analysis the portion which passes a 60-mesh B.S. test sieve and is retained on a 120-mesh B.S. test sieve.

Procedure

Place 20 g of the sample in a porcelain basin and cover with 40 ml of diluted hydrochloric acid (20% w/v, i.e. approx. 11 + 9) and boil gently for 1 h under an inverted funnel. Maintain the volume of the acid constant by adding further acid as required.

Cool, dilute with about 50 ml of water and filter through a No. 42 Whatman paper into a 250-ml volumetric flask. Wash the residue by decantation with cold water. Dilute the filtrate and washings to 250 ml and mix.

Carry out a blank determination on the acid used for the test.

(The standard test uses thioglycollic acid and describes a technique with Nessler cylinders. The following is to be preferred.)

Transfer an appropriate aliquot (usually 5–10 ml) of the diluted extract to a 100-ml volumetric flask. Add 2 ml of hydroxyammonium chloride solution (100 g/litre), 5 ml of 1:10-phenanthroline solution (10 g/litre), followed by ammonium acetate solution (approx. 10%), drop by drop, until a slight pink colour appears in the solution and then add 2 ml in excess.

Allow to stand for 15 min, dilute to 100 ml and mix.

Measure the optical density of the solution against water in 10-mm cells at 510 nm or by using a colour filter (Ilford 603) in a suitable instrument.

The colour is stable between 15 min and 75 min after the addition of the ammonium acetate solution.

Determine the ferric oxide content of the solution by reference to a calibration graph.

CHAPTER 26

DETERMINATION OF LEAD SOLUBILITY OF FRITS AND GLAZES

GENERAL

In order to conform to the Government specification for a "low-solubility" lead glaze it is necessary for the material to have a solubility under the specified conditions, of less than 5% PbO. The official regulations lay down that the test should be completed using a sulphate gravimetric method, but for most routine purposes it is better to use a chromate gravimetric finish.

The regulations demand that the attack on the glaze shall be made by shaking for 1 h and then standing for 1 h. It has been shown that unless the soluble lead is greater than about 10% PbO there is negligible error introduced by merely standing the glaze in contact with the acid for 2 h. In addition, the regulations do not specify a temperature at which the test shall be carried out, but as there is a noticeable temperature coefficient for lead chloride, it is advisable to control the temperature and 20°C has been arbitrarily chosen.

Finally, if the determination is to be carried out on a lead bisilicate frit a preliminary precipitation with hydrogen sulphide is unnecessary as the lead may be precipitated directly as chromate.

The British Ceramic Research Association has devised, for the use of the pottery industry, a test for the "intrinsic" solubility of lead bisilicate frit, wherein the frit is broken down to a close sieve range and its solubility tested under these conditions. It has been found that this intrinsic solubility bears a close relationship to the solubility of many glazes prepared from the material, thus a frit of solubility less than 1% PbO will produce glazes with a solubility of less than 5% PbO.

Thus there are two methods of dissolution of the lead, and potentially three methods of determining the lead oxide in the extract, although one of these merely eliminates the hydrogen sulphide separation. Lastly, there are two tests for lead bisilicate frits, one for intrinsic solubility and one for the official regulations. In general the simpler, non-official tests are preferable for routine work but if the solubility reaches the border line of the specification or if an "official" result is required, recourse must be made to the Government method.

Finally, if the glaze contains other elements liable to be precipitated with hydrogen sulphide, these may interfere with the chromate method and it is then advisable to use the sulphate finish.

It has been shown that atomic absorption spectrophotometry provides a very simple and elegant technique for the determination of lead in these solutions. There are, as yet, insufficient numbers of laboratories equipped with these instruments to justify inclusion of detailed procedures for this determination but the development of these in individual cases should not pose any serious problems.

Additional Reagents

Ammonium acetate
Ethanol: absolute

Prepared Reagent

Hydrogen sulphide wash solution: Add 20 ml of hydrochloric acid ($d = 1\cdot18$) to 980 ml of water and saturate with hydrogen sulphide.

Standard Solutions

Hydrochloric acid (approx. 2N)
Hydrochloric acid (0·25% HCl w/w)

PREPARATION OF SAMPLE

Almost all glazes will have been pre-ground ready for use, and will therefore be received either as fine powders, or more frequently as "slops" in water. It is necessary to dry off these slops on a steam bath and rub the resulting mass gently in a porcelain mortar to break down any lumps which may have formed and to ensure thorough mixing.

Lead bisilicate frit may be received in the ground state, in which case it is necessary to carry out the normal test as for glazes. On the other hand, the intrinsic solubility may be required and the sample will be received in the form of pieces. According to the Government method, 28 lb of frit should be taken in a random manner from the batch; this is then crushed to pass a 4-mesh B.S. test sieve and be retained on a 20-mesh B.S. test sieve. The fines should be discarded. This sample is then coned and quartered to give a few grams for the test sample.

The test sample is then dry-ground by hand to pass a 170-mesh B.S. test sieve and be retained on a 240-mesh B.S. test sieve. The fines are discarded.

PROCEDURE

Attack of the Sample

Routine Method

Weigh 0·500 g of the prepared, dried (110°C) sample on a watch glass and wash it carefully into a 500-ml Stohmann shaking flask with about 450 ml of water. Add from a burette the appropriate amount of diluted hydrochloric acid (approx. 2N) so as to make the final 500-ml volume equivalent to hydrochloric acid (0·25% w/w). [17·13 ml of hydrochloric acid (exactly 2N) is required.]

Note the exact time at which the addition is made.

Stopper the flask and shake vigorously by hand for a few seconds. Dilute to the mark with water, care being taken to wash down all the exposed particles into the acid. Allow the flask to stand for a total of 2 h from the time of the first acid addition, during which period it must be maintained at a temperature of 20 $\pm$ 1°C. (This may conveniently be done in a water bath.)

During this period prepare a Buchner funnel as follows:

Place a No. 42 Whatman paper on the grid of the funnel, cover with a $\frac{1}{4}$–$\frac{3}{4}$ in. layer of paper pulp (prepared by pulping filter paper or filter paper clippings in water, or from ashless floc), place a similar paper over the pulp and remove the surplus water under suction.

At the end of the 2 h period of standing, filter the contents of the Stohmann flask through the Buchner funnel, rejecting the first 10–15 ml of the filtrate, care being taken that the total time does not exceed 5 min, and that the filtrate is clear. If necessary, the resultant solution may be filtered through a No. 42 Whatman paper, or two of these papers folded together in an ordinary funnel.

Government Method

Weigh 0·500 g of the prepared, dried (110°C) sample on a watch glass and transfer it to a dry 500-ml Stohmann shaking flask. Wash in the last traces of the sample and fill up to the mark with hydrochloric acid (0·25% w/w). Transfer the flask to a rotary shaker and shake for 1 h. Remove the flask from the shaker and allow to stand at room temperature for 1 h.

Filtration is carried out as described above.

Precipitation of the Lead Sulphide

Accurately transfer 450 ml of the filtrate to a 600-ml beaker, warm to 80–90°C, pass a slow stream of washed hydrogen sulphide through the solution until cold, and allow to stand overnight.

Filter through a Buchner funnel with a sintered-glass mat (porosity 4) or a No. 42 Whatman paper, transferring the precipitate to the filter with a jet of cold hydrogen sulphide wash solution and wash eight times with cold hydrogen sulphide wash solution. Discard the filtrate and washings.

Determination of Lead Oxide

Chromate Method

For the determination of lead solubility of lead bisilicate frits, lead chromate may be precipitated directly in the 450-ml aliquot of solution from the extraction without preliminary precipitation of the sulphide.

Dissolve the lead sulphide precipitate through the filter with 50 ml of hot, diluted hydrochloric acid (1 + 1), and wash thoroughly with hot water. Transfer the solution and washings back to the precipitation beaker.

Evaporate the solution carefully to dryness. Cool, add 3 ml of hydrochloric acid (d = 1·18) and 100 ml of water. Heat to boiling and ensure that all the lead chloride is in solution.

Add 5 ml of potassium dichromate solution (50 g/litre), allow to cool slightly, add 20 ml of ammonium acetate solution (250 g/litre), cool and stand overnight.

Filter through a weighed crucible with a sintered-glass mat (porosity 4,) transferring the precipitate to the filter with a jet of cold water and scrubbing the beaker with a "bobby". Wash eight times with cold water. Dry at 110°C to constant weight; 2 h is usually sufficient.

Calculation—Weight of $PbCrO_4 \times 0.6906$ = weight of PbO.

If w grams is the weight of the lead chromate then the percentage of soluble PbO in the sample, x, is given by:

$$x = 153.5 \, w$$

Sulphate Method

(This method should be used if an "official" figure is required and also if it is suspected that other elements liable to be precipitated by hydrogen sulphide in acid solution are present.)

Dissolve the lead sulphide precipitate through the filter with 100 ml of hot, diluted nitric acid $(1 + 1)$ to which a few drops of bromine have been added and wash thoroughly with hot water.

Transfer the solution and washings back to the precipitation beaker, cool, cautiously add 8 ml of sulphuric acid $(d = 1.84)$ and evaporate the solution down to 3 ml on a hot plate in a fume cupboard.

Cool and add slowly, while agitating the beaker, 100 ml of water. Allow to stand for at least 4 h or preferably overnight.

Filter through an ignited and weighed silica crucible with a sintered-glass mat (porosity 4) or an asbestos-packed Gooch crucible. (If an asbestos-packed Gooch crucible is used, the asbestos pad should first be washed with water followed by eight washes with diluted sulphuric acid $(1 + 19)$ and one with ethanol (absolute). Dry and ignite over a burner to constant weight.) Transfer the precipitate to the filter with the mother liquor and scrub the beaker with a "bobby". Wash the precipitate three or four times with diluted sulphuric acid $(1 + 19)$ and finally once with ethanol (absolute). Remove as much of the alcohol as possible from the crucible by suction, then dry at 110°C in an air oven until there is no further chance of catching fire, and finally ignite at about 600°C for 5 min.

Cool and weigh, adding 0.0006 g to the weight of lead sulphate found (this being the solubility of lead sulphate in 100 ml of the sulphuric acid).

Calculation—If the corrected weight of lead sulphate is w grams, then the percentage of soluble PbO in the sample, x, is given by:

$$x = 163.6 \, w$$

CHAPTER 27

COLORIMETRY AND METHODS OF CALIBRATION

A. Colorimetry

Many of the colorimetric determinations described in the foregoing chapters can be carried out visually or with the aid of various simple types of instrument. In a few cases, for example the spectrophotometric determination of silica and alumina (Chapter 8), the accuracy obtainable by visual methods is useless and only a limited number of instruments can be used with success. Thus consideration must be given to the techniques available and the instruments with which to apply them.

The simplest technique, from the point of view of equipment and cost, is the matching of colour by eye using Nessler cylinders, but the human eye is a very poor instrument for judging colour intensities. The useful limit for this method is generally no higher than about 2% content. In addition, the actual process of carrying out the determination is usually much more protracted than when an instrument is used as it necessitates a visual comparison of colour intensities after each small addition of the requisite solution to the two Nessler cylinders.

The Lovibond Nessleriser simplifies this process by allowing visual comparison of the colour of a prepared solution with a series of tinted glasses of various intensities, but even the basic difficulties still apply. The range of the method is increased slightly to about 5% content, but the comparison is nevertheless visual.

The above methods have been largely superseded by the advent of low-priced colorimeters and absorptiometers.

Simple instruments can be purchased from about £50, but it is advisable to use an absorptiometer costing at least £80 of which the EEL absorptiometer (Fig. 27) is typical. With these instruments, wavelength selection is by means of colour filters with a transmission band several hundred ångströms wide. The effective range is thereby increased to about 5% and in favourable circumstances even higher, but such absorptiometers have the disadvantage that the calibration graphs are very often curved, and they are still not as selective as a prism or grating monochromator. The Spekker absorptiometer (Fig. 28), made by Rank Precision Industries for about £200, differs slightly from the other instruments in this category in that it utilizes a comparative system, the light passing to two photocells in order to eliminate errors due to variation of light intensity of the source. The calibration curves are still non-linear and the useful working limit of the instrument is about 5%.

When a colorimetric method has to be applied to contents greater than 5%, the use of a monochromatic type of spectrophotometer becomes essential. The Unicam SP600 spectrophotometer (Fig. 29) is the most frequently used instrument in this class as its price is only about £300, compared with about

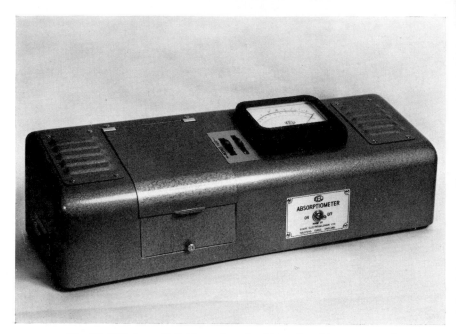

Fig. 27. EEL Long Cell Absorptiometer. Reproduced by courtesy of
Evans Electroselenium Ltd.

Fig. 28. "Spekker" Absorptiometer. Reproduced by courtesy of Rank Precision Industries.

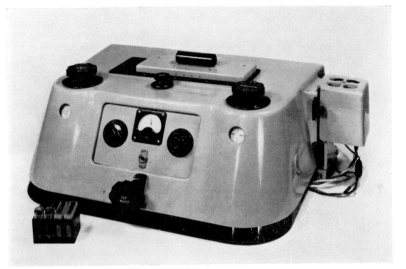

Fig. 29. Unicam SP600 Spectrophotometer. Reproduced by courtesy of
Unicam Instruments Ltd.

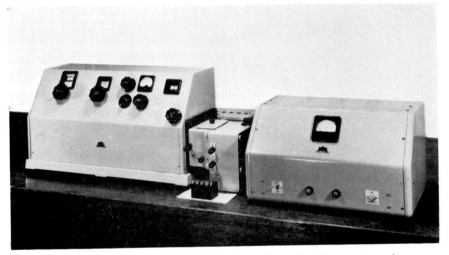

Fig. 30. Unicam SP500 Spectrophotometer. Reproduced by courtesy of
Unicam Instruments Ltd.

£900 and more for instruments such as the Unicam SP500 (Fig. 30) and the
Hilger & Watts Uvispek (Fig. 31). The former instrument is of use only in
the visual range, but with this limitation is as accurate as the others and is
particularly suitable for routine work. However, the more expensive instru-

ments have the advantage that they can be used in the ultra-violet region when such determinations are essential.

An indication of the range of the SP600 is shown by the fact that it can be used successfully for the determination of silica up to at least 95% content to an accuracy of ± 0·5%.

An interesting instrument is the Spectra (Fig. 32) made by Evans Electro-selenium Ltd. which costs only about half as much as the SP600. This uses an optical wedge for the selection of wavelengths; the instrument is not so

Fig. 31. Hilger & Watts "Uvispek". Reproduced by courtesy of Rank Precision Industries.

selective, but the spread of results is only about 50% greater than is obtained with the SP600. This means that when silica and alumina contents in the range found in aluminosilicates, and silica contents in silica bricks, are required to no greater accuracy than about ± 0·8% this instrument will be satisfactory.

From the above survey, it would appear that an absorptiometer is essential in any laboratory requiring colorimetric determinations of not more than 5% content, but if there is any possibility that the range of contents will go above this, or that determinations such as silica and alumina will at some time be needed, it is better to buy an instrument of the Unicam SP600 type at the outset.

The details in the remainder of this chapter are those which apply to the Unicam instrument. Some modifications may be necessary both to the method of working and to the technique of calibration if other instruments are used.

USE OF THE UNICAM SP600 SPECTROPHOTOMETER

As with most instruments in the laboratory, it is essential to check the performance at regular intervals. The time interval between checks will clearly

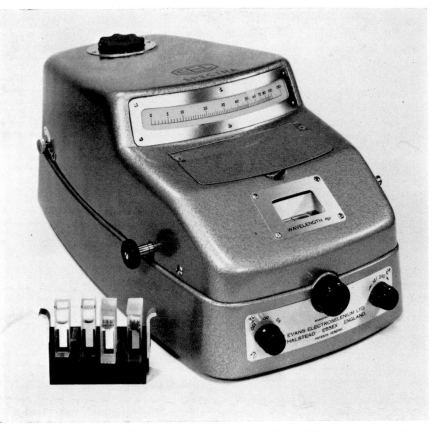

Fig. 32. EEL "Spectra". Reproduced by courtesy of Evans Electroselenium Ltd.

depend on the degree of use and the ultimate accuracy required. The wavelength drum requires checking for accuracy and in practice it is usually sufficient to do this at monthly intervals, using a didymium glass filter or, if this is not available, appropriate solutions which have sharply defined absorption bands.

Most of the calibrations described in the text are linear, so that normally it is only necessary to check two points on the graph. This is true of the methods described in Chapter 8 for the spectrophotometric analysis of high-silica materials and aluminosilicates, where details of the calibration techniques for this purpose are described separately.

For more normal use, where the blank from the reagents is negligible or is taken account of in the method, and/or where the presence of foreign ions does not disturb the colour formation, calibrations may be easily checked by reading the optical density obtained at the highest point of the previously prepared graph, at appropriate intervals. If the deviation from the calibration graph falls within the acceptable experimental error then it may be assumed that no deterioration has taken place in the instrument.

If on the other hand the variation is greater than would be expected or tolerated, then the reason should be sought. The wavelength calibration should be checked and if this is not at fault other sources of error should be explored and the matter rectified. It is not usually advisable to assume that a fresh calibration prepared at the time will be satisfactory since experience has shown that except in a very few cases the calibration graphs remain true over long periods of time; thus, when a deviation occurs the cause may well be a fault which cannot be assumed to have a constant effect.

The controls of the instrument are sensitive and must therefore be set with care; this applies particularly to the slit width control which has a rather coarse setting and for maximum accuracy must be set with care. Caution must also be applied to the siting of the instrument as it is sensitive to pressure on the bench, if this is not solid; thus the operator should not disturb the setting during use by leaning on the bench. Siting on normal laboratory benches is quite satisfactory provided that this point is noted.

No noticeable error can be detected by using the cells reversed through 180° or by placing them differently in the cell carrier, but each pair of cells should be used as a pair since the optical density of each cell, when filled with water, is slightly different. The cells should be stood in diluted chromic acid solution overnight and thoroughly rinsed out with distilled water before use. A blank, due to the difference between the units in a pair of cells, should be noted at least each day, and for work of the highest accuracy before each set of readings.

Obviously though it may seem, care must also be taken about washing the cells between uses as their awkward shape necessitates up to eight washes before it is certain that the previous solution has been removed. This can be of particular importance if additional silica, and titania are determined sequentially. The molybdate from the former determination reacts with the phosphoric acid from the latter and will produce high results in whichever determination is done last.

Calibration of the Instrument

This section gives details of the calibration of the Unicam SP600 for the determinations described in Part II, with the exception of calibrations which are specific to particular methods and in these instances details are given in the appropriate chapters. The techniques used for calibration are the simplest that can be employed for each determination, taking into account possible blanks and the effect of foreign salts. Where blanks are generally small, these are not taken into account in the calibration, but it is clear that for work of the highest accuracy, blanks should be conducted on the actual batch of reagents used and the results corrected. If foreign ions interfere with the development of the colour in any way the same amounts of salts should be present during the calibration as would be present in the determination, but where they have no effect, the calibration can be carried out without their addition.

RESIDUAL SILICA

Molybdenum Blue Method

Reagents

Prepared reagents

Ammonium molybdate (80 g/litre): Dissolve 80 g of ammonium molybdate in 1 litre of water and filter. Store in a polythene bottle and discard after 4 weeks or earlier if any appreciable deposit of molybdic acid is observed.

Stannous chloride (10 g/litre): Dissolve, by warming, 1 g of stannous chloride in 1·5 ml of hydrochloric acid ($d = 1·18$). Cool and dilute to 100 ml. This solution should be freshly prepared.

Standard solutions

Standard silica: approx. 0·5 mg SiO_2/ml.

Dilute silica standard: Dilute 20 ml of the standard silica solution to 500 ml in a volumetric flask (1 ml ≡ approx. 0·020 mg SiO_2). Prepare this solution freshly when required.

Calibration

Transfer 0, 2, 4, 6, 8 and 10 ml aliquots of the dilute silica standard solution to 100-ml volumetric flasks. Add 20, 18, 16, 14, 12 and 10 ml of water respectively. Add to each, with swirling, 5 ml of diluted hydrochloric acid (1 + 4) followed by 6 ml of ammonium molybdate solution (80 g/litre), stand for 5–10 min at a temperature of not less than 20°C and not greater than 30°C. Then add, with swirling, 45 ml of diluted hydrochloric acid (1 + 1) and stand for 10 min. Add 10 ml of stannous chloride solution (10 g/litre), dilute the solution in each flask to 100 ml and shake well.

Measure the optical density of the solutions to which silica has been added against the solution containing no silica solution, in 10-mm cells at 800 nm, or by using a colour filter (Ilford 609) in a suitable instrument. The colour is stable between 5 min and 30 min after the addition of the stannous chloride solution.

Silicomolybdate Method

Reagents

Prepared reagent

Ammonium molybdate (80 g/litre): Dissolve 80 g of ammonium molybdate in 1 litre of water and filter. Store in a polythene bottle and discard after 4 weeks or earlier if any appreciable deposit of molybdic acid is observed.

Standard solutions

Standard silica: approx. 0·5 mg SiO_2/ml.

Dilute silica standard: Dilute 40 ml of the standard silica solution to 500 ml in a volumetric flask (1 ml ≡ approx. 0·040 mg SiO_2). Prepare this solution freshly when required.

Calibration

The determination is carried out on aliquots of 0, 5, 10, 15, 20 and 25 ml of the dilute silica standard. As it is necessary to read against a control solution, two aliquots should be taken for each point and each aliquot is transferred to a 50-ml volumetric flask. Dilute the volume of each to approximately 25 ml and to each add 5 ml of diluted hydrochloric acid (1 + 4).

To one of each pair of aliquots add 5 ml of ammonium molybdate solution (80 g/litre), dilute the solution in each flask to 50 ml and shake well.

Measure the silicomolybdate colours against the appropriate control solution in 40-mm cells at 440 nm or by using a colour filter (Ilford 601) in a suitable instrument.

The colour is stable between 5 min and 15 min after the addition of the ammonium molybdate solution.

FERRIC OXIDE

Reagents

Prepared Reagents

Ammonium acetate (approx. 10%): Dilute 140 ml of acetic acid (glacial) to 2000 ml with water and add carefully 140 ml of ammonia solution ($d = 0.88$) mix, cool and adjust to approximately pH 6, either with acetic acid or ammonia solution.

1:10-phenanthroline (10 g/litre): Prepare enough solution for immediate use at a concentration of 0.1 g of 1:10-phenanthroline hydrate in 10 ml of diluted acetic acid (1 + 1).

Standard Solutions

Standard iron: 0.1 mg Fe_2O_3/ml.
Dilute iron standard: Dilute 50 ml of the standard iron solution to 500 ml in a volumetric flask (1 ml $\equiv$ 0.01 mg Fe_2O_3). Prepare this solution freshly when required.

Calibration

Transfer 0, 10, 20, 25, 40 and 50-ml aliquots of the dilute iron standard to 100-ml volumetric flasks. Add to each 2 ml of hydroxyammonium chloride solution (100 g/litre), 5 ml of 1:10-phenanthroline solution (10 g/litre) and 2 ml of ammonium acetate solution (approx. 10%). Allow to stand for 15 min, dilute the solution in each flask to 100 ml and shake well.

Measure the optical density of the solutions against water in 10-mm cells at 510 nm or by using a colour filter (Ilford 603) in a suitable instrument.

The colour is stable between 15 min and 75 min after the addition of the ammonium acetate solution.

TITANIA

Reagents

Standard Solutions

Standard titanium: 1.0 mg TiO_2/ml.

Dilute titanium standard: Dilute 20 ml of standard titanium solution to 500 ml in a volumetric flask (1 ml $\equiv$ 0·04 mg TiO_2). Prepare this solution freshly when required.

Calibration

The determination is carried out on aliquots of 0, 5, 10, 15, 20 and 25 ml of the dilute titanium standard. As it is necessary to read against a control solution, two aliquots should be taken for each point. Transfer each aliquot to a 50-ml volumetric flask and dilute to 25 ml. To each, add 10 ml of diluted phosphoric acid (2 + 3) and to one of each pair only, 10 ml of hydrogen peroxide solution (6%). Dilute the solution in each flask to 50 ml and shake well.

Measure the pertitanic acid colours against the appropriate control solution in 40 mm cells at 398 nm or by using a colour filter (Ilford 601) in a suitable instrument.

The colour is stable between 5 min and 24 h after the addition of the hydrogen peroxide solution.

MANGANESE OXIDE

Reagents

Standard Solutions

Standard manganese: 0·1 mg MnO/ml. (Prepared from standard $KMnO_4$ solution see Chapter 4).
Dilute manganese standard: Dilute 50 ml of the standard manganese solution to 500 ml in a volumetric flask (1 ml $\equiv$ 0·01 mg MnO). Prepare this solution freshly when required.

Calibration

Transfer 0, 10, 20, 25, 40 and 50-ml aliquots of the dilute manganese standard to 100-ml volumetric flasks. Dilute the solution in each flask to 100 ml and shake well.

Measure the optical density of the solutions against water in 40-mm cells at 524 nm or by using a colour filter (Ilford 604) in a suitable instrument.
NOTE: If it is desired to carry the calibration graph higher than about 0·6% this can be done by similarly calibrating with larger amounts of manganese and measuring the optical density in 10-mm cells. By this means the calibration can be extended to about 2·5% MnO.

PHOSPHORUS PENTOXIDE

Reagents

Additional Reagents

Ammonium vanadate.
Potassium dihydrogen orthophosphate.

Prepared Reagent

Ammonium molybdate (80 g/litre): Dissolve 80 g of ammonium molybdate in 1 litre of water and filter. Store in a polythene bottle and discard after 4 weeks or earlier if any appreciable deposit of molybdic acid is observed.

Standard Solution

Standard phosphate: 0·1 mg P_2O_5/ml

Calibration

The determination is carried out on aliquots of 0, 2·5, 5, 7·5, 10, and 15 ml of the standard phosphate solution. As it is necessary to read against a control solution, two aliquots should be taken for each point and each aliquot is transferred to a 100-ml volumetric flask. Dilute each to approximately 15 ml and add 10 ml of diluted nitric acid (1 + 2) and 2 drops of diluted sulphuric acid (1 + 1).

To each flask add 10 ml of ammonium vanadate solution (2·5 g/litre) and to one of each pair of aliquots add 6 ml of ammonium molybdate (80 g/litre). Dilute the solution in each flask to 100 ml and mix.

Measure the phosphovanadomolybdate colour against the appropriate control solution, in 40-mm cells at 450 nm or by using a colour filter (Ilford 601) a suitable instrument.

The colour is stable for 30 min after the addition of the ammonium molybdate solution.

NOTE: The wavelength chosen gives an optical density of about 0·7 for the equivalent of 1·5% P_2O_5. Slightly higher contents of P_2O_5 (up to about 2%) may be determined by calibrating and reading at a wavelength of 470 nm.

CHROMIUM SESQUIOXIDE

Diphenylcarbazide Method

Reagents

Additional Reagent

Diphenylcarbazide.

Prepared Reagent

Diphenylcarbazide (10 g/litre): Dissolve 0·1 g of diphenylcarbazide in 10 ml of acetone. This solution must be freshly prepared.

Standard Solutions

Standard chromium: 1·0 mg Cr_2O_3/ml.
Dilute chromium standard: Dilute 25 ml of the standard chromium solution to 1 litre in a volumetric flask (1 ml ≡ 0·025 mg Cr_2O_3). Prepare this solution freshly when required.

Calibration

Transfer 0, 1, 2, 3, 4 and 5-ml aliquots of the dilute chromium standard to 100-ml volumetric flasks. To each add 5 ml of diluted sulphuric acid (1+9) and dilute to about 90 ml. To each add 2 ml of diphenylcarbazide solution (10 g/litre), dilute to 100 ml and shake well. Allow the solutions to stand for 5 min.

Measure the optical density of the solutions against water in 10-mm cells at 540 nm or by using a colour filter (Ilford 605) in a suitable instrument.

EDTA METHOD

Reagents

Additional Reagent

  Sodium sulphite

Standard Solutions

Standard chromium: $1\cdot0$ mg Cr_2O_3/ml.
Dilute chromium standard: Dilute 50 ml of the standard chromium solution to 100 ml in a volumetric flask (1 ml $\equiv 0\cdot5$ mg Cr_2O_3). Prepare this solution freshly when required.

Calibration

Transfer 0, $5\cdot0$, $10\cdot0$, $15\cdot0$, $20\cdot0$, and $25\cdot0$ ml of the dilute chromium standard to 400-ml beakers and add 50, 45, 40, 35, 30, and 25 ml of water respectively. Add to each, 20 drops of hydrochloric acid ($d = 1\cdot18$), followed by 10 ml of sodium sulphite solution (50 g/litre) *with stirring*, and boil for 5 min. Cool to room temperature, add 10 ml of EDTA solution (50 g/litre), followed by ammonia solution ($d = 0\cdot88$), drop by drop, until the first appearance of a permanent precipitate. Dissolve this precipitate by adding 20 drops of acetic acid (1 + 1) and dilute to about 200 ml. Heat the solutions to boiling and boil for 10–15 min, cool, dilute to 250 ml in volumetric flasks and mix.

Measure the optical density of the solutions against water in 40-mm cells at 550 nm, or by using a colour filter (Ilford 605) in a suitable instrument.

NOTES ON THE USE OF THE SP600 SPECTROPHOTOMETER

Colour Development

The reagents should always be added in the correct order and into a similar volume of solution.

Care of Cells

Utmost cleanliness is essential. Daily cleaning with chromic acid followed by liberal leaching with distilled water is normally adequate.

The clear optical faces should not be handled and, after the cells have been filled, should be wiped dry with a lintless cloth or paper tissue. The optical faces should be inspected after insertion in the cell carriage to check that they are clean and dry.

"Cell blanks" should be checked at least daily, any appreciable deviation from normal indicating lack of cleanness.

Filling of Cells

Each cell should be rinsed out at least six times with the solution to be used in it. After completion of the readings the cells should be emptied and rinsed similarly with distilled water.

Operation of the Instrument

Wavelength Setting

The nature of some of the absorption bands is such that it is impossible to work on a flat portion of the curve. For this reason, if the amount of a particular constituent is on the high side, precise setting of the wavelength drum is of the greatest importance.

In this connexion it is advisable to use the left hand for operating the cell carriage to avoid accidental movement of the wavelength control.

Slit Width

This control is operated to obtain a zero reading on the blank solution with the instrument set at "Check". Again extreme care is necessary here to set the pointer exactly on the zero line, particularly as the control is very sensitive; a fractional movement of the knob causes a considerable movement of the pointer and thus results in a large potential error.

Alignment of the Cells

It is good practice, and therefore advisable, to use each cell, the same way round, in one and the same compartment of the carrier every time, to eliminate any possibility of positional errors.

Considerable errors can be introduced by failing to align the cell carrier into the cell carriage and also by failing to confirm that the carriage drops accurately into its appropriate notch.

Accuracy of Readings

It is possible to read the optical density very accurately at the lower end of the scale but, if the amount of a particular constituent is on the high side, five separate readings should be taken and the mean reading read from the graph.

CHAPTER 28

RATIONAL ANALYSIS

Although methods have been suggested from time to time for separating the mineralogical compounds of clays, these have not proved particularly satisfactory. Adequate estimates of the chief components can be made, however, by calculation from the ultimate analysis on the basis of limited assumptions. These so-called proximate analyses, calculated rational analyses or, preferably, calculated mineralogical compositions can provide the manufacturer with a sufficiently adequate picture for most practical purposes.

Originally the method was applied by the potter to clays, but the practice adopted by mineralogists of introducing "norms" in estimating the mineralogical content of rock samples has been adapted to ceramic purposes notably in the assessment of feldspathic fluxes, Cornish stone, nepheline syenite, etc.

The calculated mineralogical composition of clays has followed two main lines. The older method was to consider all basic oxides as occurring as orthoclase feldspar, $K_2O.Al_2O_3.6SiO_2$, so that the oxides Na_2O,CaO,MgO were all regarded as K_2O. By multiplying by the appropriate factor, the amount of "ideal" feldspar could be estimated and from this, the amount of alumina and silica present as feldspar could be calculated. Any residual alumina was regarded as being present as kaolinite, $Al_2O_3.2SiO_2. 2H_2O$. From the total kaolinite content, the amount of silica and water of constitution in the clay could be estimated. Any residual silica was regarded as quartz and any residual loss on ignition as carbonaceous matter.

This feldspathic convention was later modified to allow lime and magnesia to appear as carbonates with corresponding adjustment to the carbonaceous matter. On occasions when the alumina present was in excess of the silica necessary for the clay and feldspar compositions the excess was regarded as corundum.

The procedure for calculating mineralogical composition on this revised feldspar convention was:

(1) Total alkalis $\times$ 5·9 = "ideal" feldspar.
(2) "Ideal feldspar $\times$ 0·183 = Al_2O_3 in "ideal" feldspar.
(3) "Ideal" feldspar $\times$ 0·647 = SiO_2 in "ideal" feldspar.
(4) Residual $Al_2O_3 \times$ 2·53 = clay substance (kaolinite).
(5) Kaolinite $\times$ 0·465 = SiO_2 in kaolinite.
(6) Total SiO_2 — (SiO_2 in kaolinite + SiO_2 in feldspar) = quartz.
(7) MgO $\times$ 2·1 = magnesium carbonate.
(8) CaO $\times$ 1·78 = calcium carbonate.

Recent evidence has shown that in many instances much of the magnesium and iron are present in the clay lattice and that some of the lime may be present as exchangeable cations associated with the clay. Moreover, mica is almost universally present and feldspar rarely, if ever, in sedimentary clays.

The modern trend in calculating the "mineralogical" composition is to use the so-called "mica convention" which deals with the alkalis, alumina, silica and ignition loss, and to list the remaining elements as oxides unless there is direct evidence of the form in which they occur as, for instance, with TiO_2 which is usually present as anatase.

With this calculation the alkalis are calculated separately as "paragonite" (soda mica) and "muscovite" (potash mica), the remaining alumina appears as kaolinite, the remaining silica as "quartz" and any excess loss on ignition as "carbonaceous matter". The factors are as follows:

Quartz $\quad = SiO_2 - 1 \cdot 178Al_2O_3$
Muscovite $\quad = K_2O \times 8 \cdot 46$
Paragonite $\quad = Na_2O \times 12 \cdot 33$
$\quad SiO_2$ in micas $\quad = 3 \cdot 83 \times K_2O + 5 \cdot 81 \times Na_2O = a$
$\quad \therefore SiO_2$ in kaolinite $= 1 \cdot 178Al_2O_3 - a = b$
and Kaolinite $\quad = 2 \cdot 15 \times b$

Bound water $\quad = c = 0 \cdot 3534Al_2O_3 - \dfrac{a}{5}$

Carbonaceous matter $\quad = $ loss on ignition $- c$

If the free quartz is negative, free alumina is considered to be present.

Free alumina $\quad = Al_2O_3 - \dfrac{SiO_2}{1 \cdot 178}$

$\quad SiO_2$ in kaolinite $\quad = SiO_2 - a = d$
Kaolinite $\quad = 2 \cdot 15d$

Bound water $\quad = e = \dfrac{3}{10} SiO_2 - \dfrac{a}{5}$

Carbonaceous matter $\quad = $ loss on ignition $- e$

With this calculation it is the exception to find negative values for the carbonaceous matter; with the feldspar convention such occasions are more frequent, even when the clay is demonstrably carbonaceous. This fact adds further credence to the mica calculation although it is admittedly imperfect.

SOME SPECIMEN ANALYSES

The following section contains analyses of a representative selection of materials which have been analysed in the laboratories of the British Ceramic Research Association.

It is hoped that these will act as a guide, to analysts unfamiliar with a particular material, as to the constituents requiring determination.

It is appreciated that these tables may not be comprehensive, but the examples chosen are generally representative of large numbers analysed by the Association.

All values are expressed as percentages. For convenience of presentation, the materials are grouped with one list of constituents serving for each group. As a result, the constituents listed under various materials include some that would not normally be sought in those materials, Such constituents were, of course, not determined and are marked "n.d." accordingly. Constituents detected qualitatively, but in insufficient amount to be determined quantitatively, are marked "tr.", i.e. "trace".

Those samples which are coded B.C.S. followed by a number are available from the Bureau of Analysed Samples Limited, Newham Hall, Newby, Middlesbrough.

	Aluminous Cement	Amblygonite	Andalusite
SiO_2	33·46	7·1	43·35
TiO_2	1·46	0·07	0·39
Al_2O_3	59·78	32·2	52·92
Fe_2O_3	1·26	0·04	1·60
CaO	0·09	0·3	0·32
MgO	0·23	0·5	0·19
Na_2O	0·06	1·1	0·21
K_2O	0·88	0·3	0·41
Li_2O	0·03	7·3	0·01
MnO	n.d.	n.d.	n.d.
P_2O_5	n.d.	42·8	n.d.
F	n.d.	3·4	n.d.
SO_3	n.d.	0·06	n.d.
Loss	2·72	5·7	0·92
	99·97	100·87	100·32

Oxygen equivalent of F   1·4

99·5

	Anorthite	Ash: boiler	Ash: coal	Ball clay: black	Ball clay: blue	Ball clay: white	Ball clay: siliceous
SiO_2	45·10	17·85	39·66	59·93	49·12	57·07	76·98
TiO_2	0·14	0·50	1·15	1·21	0·11	0·74	1·15
Al_2O_3	33·75	9·48	23·15	26·03	35·73	29·33	14·34
Fe_2O_3	1·20	13·97*	11·25*	1·69	0·56	1·19	1·35
CaO	17·01	6·85	6·08	0·15	0·18	0·19	0·17
MgO	0·72	2·06	1·59	0·52	0·24	0·36	0·26
Na_2O	0·14	0·42	1·92	0·14	0·39	0·10	0·20
K_2O	1·62	0·82	1·44	3·42	1·46	1·28	1·66
Li_2O	0·05	n.d.	n.d.	n.d.	n.d.	n.d.	n.d.
MnO	0·01	0·45	0·09	n.d.	n.d.	n.d.	n.d.
SO_3	n.d.	1·98†	0·40†	n.d.	n.d.	n.d.	n.d.
Loss	0·31	46·03	13·67	6·87	11·93	9·53	4·30
	100·05	100·41	100·40	99·96	99·72	99·79	100·41

*Total iron as Fe_2O_3.   † After ignition.

	Barium carbonate	Barium sulphate	Barytes	Bauxite	Bentonite	Body: china
SiO_2	0·02	0·16	18·65	5·93	62·13	30·72
TiO_2	<0·01	<0·01	0·05	3·14	0·25	<0·01
Al_2O_3	0·06	<0·01	0·23	89·04	21·40	13·44
Fe_2O_3	0·01	0·01	0·08	1·74	3·51	0·50
CaO	0·24	0·04	24·28	0·10	1·03	25·84
MgO	0·15	0·03	0·05	0·03	3·29	0·04
Na_2O	0·29	0·08	0·10	0·01	0·92	0·26
K_2O	0·03	0·05	0·06	0·02	0·44	1·28
Li_2O	<0·01	<0·01	n.d.	<0·01	n.d.	n.d.
BaO	75·95	64·46	30·65	n.d.	n.d.	n.d.
SrO	tr.	tr.	n.d.	n.d.	n.d.	n.d.
P_2O_5	<0·01	<0·01	n.d.	n.d.	n.d.	22·89
SO_3	0·31*	33·55*	17·20*	n.d.	n.d.	n.d.
F	n.d.	n.d.	14·26	n.d.	n.d.	n.d.
Loss	n.d.	n.d.	0·40	0·10	6·87	4·66
CO_2	22·16†	<0·01	n.d.	n.d.	n.d.	n.d.
C	<0·01	<0·01	n.d.	n.d.	n.d.	n.d.
H_2O	0·86‡	1·29‡	n.d.	n.d.	n.d.	n.d.
	100·08	99·67	106·01	100·11	99·84	99·63

$$\text{Oxygen equivalent of F } \underline{6·00}$$
$$\underline{100·01}$$

* After ignition.   † Total.   ‡ Combined.

	Body: earthenware	Body: fireclay	Bone ash	Cement	China clay
SiO_2	69·78	56·95	0·30	20·46	46·62
TiO_2	0·30	1·07	<0·01	0·39	0·07
Al_2O_3	18·93	29·17	0·04	4·79	38·31
Fe_2O_3	0·50	2·18	0·12	3·25	0·38
CaO	1·19	0·43	54·16	64·86	0·33
MgO	0·24	0·51	1·23	1·51	0·25
Na_2O	0·68	0·21	0·28	0·56	0·30
K_2O	1·66	1·86	<0·01	0·27	0·68
Li_2O	n.d.	n.d.	n.d.	0·05	n.d.
MnO	n.d.	n.d.	n.d.	0·09	n.d.
P_2O_5	n.d.	n.d.	41·12	n.d.	n.d.
SO_3	n.d.	n.d.	0·19*	2·47*	n.d.
Loss	6·45	7·99	1·35	1·33	13·20
CO_2	n.d.	n.d.	0·65	n.d.	n.d.
Cl	n.d.	n.d	0·04	n.d.	n.d.
	99·73	100·37	99·48	100·03	100·14

*After ignition.

L

	BCS 308 Grecian chrome ore	Chrome- magnesite refractory	Diatomite	BCS 368 Dolomite
SiO_2	4·25	4·21	91·67	0·92
TiO_2	0·12	0·16	0·18	<0·01
Al_2O_3	19·4	8·74	2·82	0·17
Fe_2O_3	17·0*	11·85	0·91	0·23
CaO	0·34	0·77	0·16	30·8
MgO	16·4	37·19	0·44	20·9
Na_2O	0·04	0·07	0·74	<0·01
K_2O	0·01	0·08	0·68	<0·01
Li_2O	0·03	0·04	n.d.	<0·01
MnO	0·12	0·10	n.d.	0·06
Cr_2O_3	41·5	36·79	n.d.	<0·01
Loss	0·80	0·28	2·50	46·7
	100·01	100·28	100·10	99·78

*Total iron as Fe_2O_3.

	Feldspar (Potash)	Feldspar (Soda)	BCS 315 Firebrick	Fireclay	Fireclay
SiO_2	66·93	71·95	51·2	44·32	66·45
TiO_2	0·02	0·34	1·23	1·23	1·42
Al_2O_3	17·64	15·65	42·4	37·03	19·69
Fe_2O_3	0·30	0·13	3·01	1·94	2·44
CaO	0·47	1·14	0·43	0·30	0·22
MgO	0·13	0·03	0·57	0·62	0·63
Na_2O	2·88	8·47	0·13	0·20	0·34
K_2O	10·8	0·72	0·52	0·32]	1·64
Li_2O	n.d.	n.d.	0·08	n.d.	n.d.
MnO	n.d.	n.d.	0·02	n.d.	0·02
Loss	0·80	1·47	0·26	14·24	7·27
	99·97	99·90	99·85	100·20	100·12

	Flint	Fluorspar	Fluorspar	Frit: borax	Frit: lead bisilicate	Frit: zircon	Ganister
SiO$_2$	98·17	1·02	0·78	49·03	31·14	54·10	98·56
TiO$_2$	0·04	0·13	0·02	0·06	<0·01	0·07	0·25
Al$_2$O$_3$	0·82	0·95	<0·01	7·81	3·02	8·63	0·74
Fe$_2$O$_3$	0·02	1·05	0·22	0·26	0·08	0·26	0·03
CaO	0·52	61·96	50·36	16·41	0·06	7·36	0·03
MgO	0·03	0·49	<0·01	0·14	<0·01	0·18	0·06
Na$_2$O	0·02	0·10	0·09	9·48	0·06	4·33	0·02
K$_2$O	0·09	0·12	0·04	1·17	0·19	1·80	0·02
Li$_2$O	n.d.	n.d.	<0·01	0·05	n.d.	0·10	n.d.
MnO	n.d.	0·01	n.d.	n.d.	n.d.	n.d.	n.d.
Cr$_2$O$_3$	n.d.	0·07	n.d.	n.d.	n.d.	n.d.	n.d.
B$_2$O$_3$	n.d.	n.d.	n.d.	14·88	n.d.	12·96	n.d.
PbO	n.d.	n.d.	n.d.	n.d.	65·05	n.d.	n.d.
ZnO	n.d.	3·88	n.d.	n.d.	n.d.	3.05	n.d.
BaO	n.d.	0·70	18·87	n.d.	n.d.	n.d.	n.d.
ZrO$_2$	n.d.	n.d.	n.d.	n.d.	n.d.	7·08	n.d.
P$_2$O$_5$	n.d.	0·15	n.d.	0·12	n.d.	n.d.	n.d.
SO$_3$	n.d.	1·05*	10·10*	n.d.	n.d.	n.d.	n.d.
F	n.d.	37·24	33·50	n.d.	n.d.	n.d.	n.d.
Loss	0·39	6·43	0·15	0·18	0·14	0·18	0·27
	100·10	115·35	114·13	99·59	99·74	100·10	99·98
Oxygen equiv. of F		15·68	14·10				
		99·67	100·03				

*After ignition.

	Glaze: Soda-lime -silica	Glaze: leadless	Glaze: leadless	Glaze: low sol.	Granite
SiO$_2$	73·06	60·29	52·82	49·62	73·24
TiO$_2$	0·10	0·07	0·05	0·04	0·42
Al$_2$O$_3$	0·95	8·20	14·80	9·32	13·26
Fe$_2$O$_3$	0·28	0·16	0·35	0·26	2·50
CaO	8·97	6·76	9·03	7·73	2·23
MgO	3·10	<0·01	0·22	0·08	1·06
Na$_2$O	12·68	5·13	5·61	5·00	3·48
K$_2$O	0·57	0·79	4·22	1·35	2·86
Li$_2$O	0·05	n.d.	0·06	0·02	n.d.
MnO	n.d.	n.d.	n.d.	n.d.	n.d.
B$_2$O$_3$	n.d.	11·76	10·34	7·12	n.d.
PbO	n.d.	1·11	n.d.	17·34	n.d.
ZnO	n.d.	n.d.	n.d.	n.d.	n.d.
BaO	n.d.	4·22	1·63	n.d.	n.d.
SnO$_2$	n.d.	n.d.	n.d.	n.d.	n.d.
SO$_3$	n.d.	n.d.	n.d.	n.d.	n.d.
Loss	0·11	1·13	1·20	1·66	1·07
	99·87	99·62	100·33	99·54	100·12

	Kyanite	Lime-stone	Lime-stone	Magne-site	*Magne-site	*Magne-site	BCS 369 *Magnesite -chrome refractory
SiO$_2$	40·78	9·62	0·59	0·59	2·49	0·97	2·58
TiO$_2$	0·84	0·18	<0·01	0·04	0·04	0·02	0·14
Al$_2$O$_3$	57·16	1·65	0·38	0·10	0·85	0·99	14·7
Fe$_2$O$_3$	0·28	1·44	0·12	0·18	1·77	5·44	10·3
Cr$_2$O$_3$	n.d.	n.d.	n.d.	n.d.	0·07	0·75	17·2
CaO	0·16	47·44	54·75	1·28	2·32	1·76	1·17
MgO	0·02	0·98	0·20	97·70	92·20	89·57	53·5
Na$_2$O	0·14	0·12	0·03	0·05	0·03	0·03	0·05
K$_2$O	0·10	0·40	0·05	0·02	0·02	0·01	0·03
Li$_2$O	n.d.	n.d.	<0·01	n.d.	<0·01	<0·01	0·03
MnO	n.d.	n.d.	n.d.	<0·01	0·10	0·12	0·11
Loss	0·60	37·87	43·45	0·37	—	—	—
	100·08	99·70	99·57	100·33	99·89	99·66	99·81

*Results reported on ignited basis (1000°C).

	Marl	Molochite	Nepheline syenite	Petalite	Plaster
SiO$_2$	59·62	53·91	49·53	75·39	2·80
TiO$_2$	1·07	0·15	0·35	<0·01	0·12
Al$_2$O$_3$	19·86	42·12	24·00	18·88	0·44
Fe$_2$O$_3$	8·94	0·84	2·85	0·04	0·20
CaO	0·52	0·25	3·10	0·06	36·66
MgO	0·81	0·22	0·28	0·01	0·72
Na$_2$O	0·11	0·56	8·33	0·31	0·15
K$_2$O	1·81	1·60	8·34	0·28	0·24
Li$_2$O	0·02	n.d.	n.d.	4·10	n.d.
MnO	0·04	n.d.	0·10	n.d.	n.d.
P$_2$O$_5$	n.d.	n.d.	n.d.	0·22	n.d.
CO$_2$	n.d.	n.d.	n.d.	n.d.	0·84
SO$_3$	n.d.	n.d.	n.d.	n.d.	51·54
Loss	7·31	0·32	3·05	0·44	n.d.
H$_2$O	n.d.	n.d.	n.d.	n.d.	5·95*
	100·11	99·97	99·93	99·73	99·66

*Combined.

	Plaster	Quartz	Rutile	Sand: red	Sand: silica	BCS 313 Sand: high-purity silica	Sandstone
SiO_2	6·19	99·75	0·60	85·38	99·30	99·6	92·12
TiO_2	0·48	0·01	97·94	0·42	0·17	0·022	0·72
Al_2O_3	3·98	0·13	0·08	7·10	0·22	0·16	3·68
Fe_2O_3	0·16	0·02	1·03	1·69	0·18	0·030	0·75
CaO	32·95	0·01	0·02	0·18	0·07	0·02	0·26
MgO	0·27	0·01	0·02	0·46	0·02	<0·01	0·13
Na_2O	0·36	0·02	0·18	0·40	<0·01	<0·01	0·24
K_2O	0·48	<0·01	0·18	2·64	0·02	0·04	0·66
Li_2O	n.d.	n.d.	n.d.	n.d.	n.d.	<0·01	n.d.
Cr_2O_3	n.d.	n.d.	n.d.	n.d.	n.d.	<0·001	n.d.
MnO	n.d.	n.d.	0·03	0·01	n.d.	<0·005	0·02
SO_3	28·36*	n.d.	n.d.	n.d.	n.d.	n.d.	n.d.
Loss	26·46	0·04	0·11	1·51	0·06	0·13	1·25
	99·69	99·99	100·19	99·79	100·04	100·00	99·83

*After ignition.

	Sepiolite	Serpentine	BCS 314 Silica brick	Siliceous brick	Silicon carbide*	Silicon carbide†	Sillimanite
SiO_2	51·16	59·69	96·2	87·70	129·32	24·37‡	40·30
TiO_2	0·35	0·11	0·19	0·40	0·16	0·16	1·48
Al_2O_3	12·22	0·84	0·77	9·92	3·98	3·98	55·08
Fe_2O_3	5·19	0·72	0·53	1·01	0·32	0·32	0·98
CaO	0·54	0·02	1·81	0·12	0·16	0·16	0·48
MgO	18·84	30·60	0·05	0·18	0·06	0·06	0·32
Na_2O	0·49	0·04	0·05	0·08	0·52	0·52	0·46
K_2O	2·55	0·04	0·09	0·76	0·64	0·64	0·76
Li_2O	0·03	n.d.	0·01	0·01	n.d.	n.d.	n.d.
MnO	0·06	<0·01	<0·01	n.d.	n.d.	n.d.	n.d.
FeO	n.d.	2·23	n.d.	n.d.	n.d.	n.d.	n.d.
Cr_2O_3	n.d.	0·29	n.d.	n.d.	n.d.	n.d.	n.d.
C	n.d.	n.d.	n.d.	n.d.	20·97	(SiC) 70·01	n.d.
Loss	8·76	5·14	0·11	0·02	0·04	0·04	0·18
	100·19	99·72	99·81	100·20	156·17	100·26	100·04

Oxygen equivalent of C 55·92

100·25

* Actual analysis.   † Calculated analysis from the same results.
‡ Free silica, after calculating all the carbon to silicon carbide.

	Slag: coke oven	Slag: fuel oil	Steatite: talc, soap-stone, etc.	Steatite: talc, soap-stone, etc.	Stone: Cornish
SiO_2	55·80	45·42	44·86	61·08	72·29
TiO_2	0·77	0·94	0·09	0·08	0·17
Al_2O_3	11·24	25·56	0·99	0·91	15·30
Fe_2O_3	24·20*	2·39*	4·94	0·35	0·27
CaO	3·93	2·62	1·75	0·25	1·74
MgO	2·17	1·07	35·40	31·13	0·19
Na_2O	0·51	8·60	0·14	0·34	3·42
K_2O	1·88	1·17	0·02	0·12	4·10
Li_2O	0·07	0·08	n.d.	n.d.	0·04
MnO	0·09	0·05	0·04	n.d.	n.d.
V_2O_5	n.d.	12·50	n.d.	n.d.	n.d.
Cr_2O_3	n.d.	tr.	0·38	n.d.	n.d.
BaO	n.d.	tr.	n.d.	n.d.	n.d.
P_2O_5	n.d.	n.d.	n.d.	n.d.	1·03
F	n.d.	n.d.	n.d.	n.d.	0·90†
Loss	−0·28‡	−0·43‡	11·17	5·44	1·46
	100·38	99·97	99·78	99·70	100·01

* Total iron as Fe_2O_3.   † Fluorine not included in the total.
‡ Gain on ignition.

	Tin oxide	Vermiculite	Wollastonite	Zircon batt	Zircon brick
SiO_2	0·15	42·90	52·10	44·66	33·96
TiO_2	<0·01	0·93	<0·01	0·34	0·25
Al_2O_3	0·07	10·29	0·51	30·26	0·18
Fe_2O_3	0·05	6·22	0·05	0·84	0·13
CaO	0·08	3·77	40·49	0·06	0·30
MgO	0·06	31·06	0·61	5·64	0·04
Na_2O	0·02	0·08	0·05	0·25	0·02
K_2O	<0·01	4·16	0·05	1·28	0·02
Li_2O	n.d.	n.d.	<0·01	n.d.	n.d.
MnO	n.d.	<0·01	n.d.	n.d.	n.d.
PbO	0·12	n.d.	n.d.	n.d.	n.d.
SnO_2	99·07	n.d.	n.d.	n.d.	n.d.
Cr_2O_3	n.d.	0·17	n.d.	n.d.	n.d.
ZrO_2	n.d.	n.d.	n.d.	16·42	65·18
SO_3	n.d.	0·15*	n.d.	n.d.	n.d.
Loss	0·10	0·60	6·17	0·08	0·04
	99·72	100·33	100·03	99·83	100·12

* After ignition.

INDEX